D0946169

The New Solar Home Book

Bruce Anderson
with Michael Riordan

Brick House Publishing Company
Andover, Massachusetts 01810

Acknowledgements

The first edition of this book, titled *The Solar Home Book*, was based on Bruce Anderson's master's thesis, "Solar Energy and Shelter Design", for the School of Architecture at M.I.T. His manuscript was revised for book publication by Michael Riordan. This edition was produced by the staff of Cheshire Books under the direction of Linda Goodman. Illustrations were by Edward A. Wong.

Revisions to bring the book up to date for the second edition were done by Jennifer Adams, a designer with The Write Design and former engineering editor of *Solar Age* magazine (now *Progressive Builder*). Additional illustrations were prepared by ANCO of Boston.

Publication of both editions has been financed through the efforts of Richard Katzenberg.

Library of Congress Cataloging-in-Publication Data

Anderson, Bruce, 1947-
   The new solar home book.

   Rev. ed. of: The solar home book. c1976.
   Includes index.
   1. Solar houses. 2. Solar energy. I. Riordan,
Michael. II. Anderson, Bruce, 1947-   . Solar home
book. III. Title.
TH7413.A53 **1987**   697'.78   86-23214
ISBN 0-931790-70-0 (pbk.)

Foreword

For generations, Americans have viewed cheap and plentiful energy as their birthright. Coal, oil or gas have always been abundantly available to heat our homes, power our automobiles, and fuel our industries. But just as the supply of these fossil fuels begins to dwindle and we look to the atom for salvation, we are beginning to perceive the environmental havoc being wrought by our indiscriminate use of energy. Our urban and suburban skies are choked with smog; our rivers and shores are streaked with oil; even the food we eat and the water we drink are suspect. And while promising us temporary relief from energy starvation, nuclear power threatens a new round of pollution whose severity is still a matter of speculation.

The residential use of solar energy is one step toward reversing this trend. By using the sun to heat and cool our homes, we can begin to halt our growing dependence on energy sources that are polluting the environment and rising in cost. The twin crises of energy shortage and environmental degradation occur because we have relied on concentrated forms of energy imported from afar. We had little say in the method of energy production and accepted its by-products just as we grasped for its benefits. But solar energy can be collected right in the home, and we can be far wiser in its distribution and use.

Unlike nuclear power, solar energy produces no lethal radiation or radioactive wastes. Its generation is not centralized and hence not open to sabotage or blackmail. Unlike oil, the sun doesn't blacken our beaches or darken our skies. Nor does it lend itself to foreign boycott or corporate intrigue. Unlike coal, the use of solar energy doesn't ravage our rural landscapes with strip mining or our urban atmospheres with soot and sulphurous fumes.

Universal solar heating and cooling could ease fuel shortages and environmental pollution substantially. Almost 15 percent of the energy consumed in the United States goes for home heating, cooling, and water heating. If the sun could provide two thirds of these needs, it would reduce the national consumption of non-renewable fuels by 10 percent and world consumption by more than 3 percent. National and global pollution would drop by similar amounts.

But solar energy has the drawback of being diffuse. Rather than being mined or drilled at a few scattered places, it falls thinly and fairly evenly across the globe. The sun respects no human boundaries and is available to all. Governments and industries accustomed to concentrated energy supplies are ill-equipped, by reason of economic constraints or philosophical prejudices, to harness this gentle source of energy. These institutions are far more interested in forms

Foreword

of energy that lend themselves to centralization and control. Hence the United States government spends billions for nuclear power while solar energy is just a subject for study—a future possibility, maybe, but not right now.

This book speaks to the men and women who cannot wait for a hesitant government to "announce" a new solar age. We can begin to fight energy shortages and environmental pollution in our own homes and surroundings. Solar heating and cooling are feasible *today*—not at some nebulous future date. The solar energy falling on the walls and roof of a home during winter is several times the amount of energy needed to heat it. All it takes to harness this abundant supply is the combination of ingenuity, economy and husbandry that has been the American ideal since the days of Franklin and Thoreau.

Bruce Anderson
Harrisville, New Hampshire

Michael Riordan
Menlo Park, California

Contents

Contents

Contents

Introduction

Now in houses with a south aspect, the sun's rays penetrate into the porticoes in winter, but in summer the path of the sun is right over our heads and above the roof, so that there is shade. If, then, this is the best arrangement, we should build the south side loftier to get the winter sun and the north side lower to keep out the cold winds.

Socrates, as quoted by
Xenophon in *Memorabilia*

The design of human shelter has often reflected an understanding of the sun's power. Primitive shelters in tropical areas have broad thatched roofs that provide shade from the scorching midday sun and keep out frequent rains. The open walls of these structures allow cooling breezes to carry away accumulated heat and moisture. In the American southwest, Pueblo Indians built thick adobe walls and roofs that kept the interiors cool during the day by absorbing the sun's rays. By the time the cold desert night rolled around, the absorbed heat had penetrated the living quarters to warm the inhabitants. Communal buildings faced south or southeast to absorb as much of the winter sun as possible.

Even the shelters of more advanced civilizations have been designed to take advantage of the sun. The entire Meso-American city of Teotihuacan, the size of ancient Rome, was laid out on a grid facing 15 degrees west of south. Early New England houses had masonry filled walls and compact layouts to minimize heat loss during frigid winter months. The kitchen, with its constantly burning wood stove, was located on the north side of the house to permit the other rooms to occupy the prime southern exposure. Only in the present century, with abundant supplies of cheap fossil fuels available, has the sun been ignored in building design.

Serious technical investigations into the use of the sun to heat homes began in 1939, when the Massachusetts Institute of Technology built its first solar house. For the first time, *solar collectors* placed on the roof gathered sunlight for interior heating. By 1960, more than a dozen structures had been built to use modern methods of harnessing the sun's energy.

During the 1970s, following the Arab oil embargo, thousands of solar homes were built. Hundreds of manufacturers produced solar collectors, and the sun's energy was used to heat domestic water as well. But the steep rise in crude oil prices also triggered conservation on a scale that dramatically cut worldwide oil consumption, forcing crude oil prices back down. Widespread popular interest in energy subsided momentarily, but did leave behind a legacy of real progress in the uses of renewable energy.

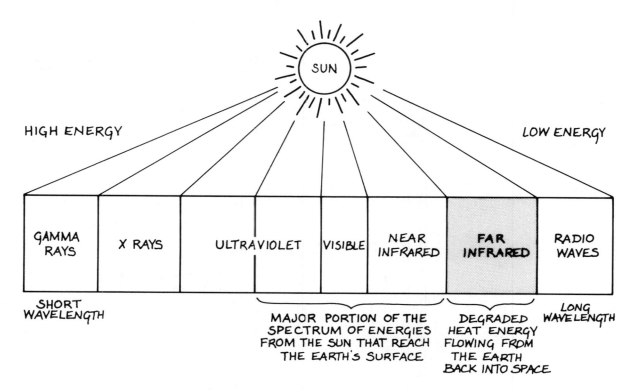

HIGH ENERGY   LOW ENERGY

| GAMMA RAYS | X RAYS | ULTRAVIOLET | VISIBLE | NEAR INFRARED | FAR INFRARED | RADIO WAVES |

SHORT WAVELENGTH

MAJOR PORTION OF THE SPECTRUM OF ENERGIES FROM THE SUN THAT REACH THE EARTH'S SURFACE

DEGRADED HEAT ENERGY FLOWING FROM THE EARTH BACK INTO SPACE

LONG WAVELENGTH

The electromagnetic spectrum. (Miller, *Living in the Environment*. Wadsworth.)

SOLAR AND HEAT BASICS

Most of the solar energy reaching us comes in the form of visible light and *infrared* rays. These two forms of radiation differ only in their wavelengths. When they strike an object, part of the radiation is absorbed and transformed into an equivalent amount of heat energy. Heat is simply the motion of atoms and molecules in an object. It is stored in the material itself or *conducted* to surrounding materials, warming them in turn. Heat can also be carried off by air and water flowing past these warm materials, in what is called *convection* heat flow.

That a material can be heated by the sun is obvious to anyone who has walked barefoot over a sun-baked pavement. What may not be so obvious is that the pavement also *radiates* some of the heat energy away in the form of infrared rays. You can feel this *thermal radiation* by putting your hand near an iron poker after it has been heated in a fireplace. It is this radiation of energy back into space that keeps the earth from overheating and frying us to a crisp.

The amount of solar energy reaching the earth's surface is enormous. It frequently exceeds 200 Btu per hour on a square foot of surface, enough to power a 60–watt light bulb if all the solar energy could be converted to electricity. But the technology of solar electricity is in its infancy; we are fortunate if we can convert even 15 percent. On the other hand, efficiencies of 60 percent are not unreasonable for the conversion of solar energy into heat for a house. The energy falling on a house during the winter is generally several times what is needed inside, so the sun can provide a substantial fraction of its annual needs.

Glass is the "miracle" substance that makes solar heating possible. Glass transmits visible light but not thermal radiation. You can prove this to yourself by sitting in front of a blazing fire. Your face becomes unbearably hot if you sit too close. But what happens if you place a pane of glass in front of your face? You can still *see* the fire but your face is not nearly as hot as before. The longwave infrared rays carrying most of the fire's radiant energy are absorbed by the glass, while the shortwave visible

2

rays penetrate to your eyes. In the same way, once sunlight passes through a window and is transformed into heat energy inside, this energy cannot be radiated directly back outside. This phenomenon, known as the *greenhouse effect,* is responsible for the hot, stuffy air in the car you left in the sun after the doors locked and the windows rolled up. Other transparent materials, particularly plastics, also absorb this thermal radiation, but none quite so well as glass.

The basic principles of solar collection for home heating and cooling are embodied in the greenhouse. The sun's rays pass through the glass or transparent plastic *glazing* and are absorbed by a dark surface. The heat produced cannot escape readily because thermal radiation and warm air currents are trapped by the glazing. The accumulated solar heat is then transported to the living quarters or stored.

There is often an overabundance of solar energy when it is not needed, and none at all when it is most in demand. Some means is required to *store* the collected solar heat for use at night or during extended periods of cloudiness. Any material absorbs heat as its temperature rises and releases heat as its temperature falls. The objects inside a house—the walls, ceilings, floors, and even furniture—can serve as heat storage devices.

Measurement of Heat and Solar Energy

There are two basic types of measurement used to describe heat energy—temperature and quantity. Temperature is a measure of the average vibrational energy of molecules. For example, the molecules in water at 40°C (degrees Centigrade) are vibrating more rapidly than molecules in water at 10°C. Heat quantity is determined both by how rapidly molecules are vibrating and by how many molecules there are. For example, it takes a much larger quantity of heat to raise a swimming pool to 40°C than to raise a kettle of water to 40°C, even though the temperature is the same in both.

In the English system of measurement, the unit of heat quantity is the British Thermal Unit, or Btu, the amount of heat needed to raise one pound of water one degree Fahrenheit (°F). In the metric system, the unit of heat quantity is the calorie, or cal, the amount of heat required to raise one gram of water one degree Centigrade. One Btu is equivalent to about 252 cal. It takes the same quantity of heat, 100 Btu or 25,200 cal, to heat 100 pounds of water 1°F as it does to heat 10 pounds of water 10°F.

Heat is one form of energy and sunlight is another—radiant energy. An important characteristic of energy is that it is never lost—energy may change from one form to another, but it never disappears. Thus we can describe the amount of solar energy striking a surface in terms of an equivalent amount of heat. We measure the solar energy striking a surface in a given time period in units of Btu/ft^2/hr or cal/cm^2/min. Outside the earth's atmosphere, for example, solar energy strikes at the average rate of 429 Btu/ft^2/hr or 1.94 cal/cm^2/min.

The radiant energy reaching us from the sun has a distribution of wavelengths (or colors). We describe these wavelengths in units of microns, or millionths of a meter. The wavelength distribution of solar energy striking the earth's atmosphere and reaching the ground is shown in the accompanying chart.

About half of the solar radiation reaching the ground falls in the visible range, 0.4 to 0.7 microns. Most of the radiation in the ultraviolet range, with wavelengths below 0.4 microns, is absorbed in the upper atmosphere. A substantial portion of the infrared radiation, with wavelengths greater than 0.7 microns, reaches the earth's surface. A warm body emits even longer wave infrared radiation. Since glass transmits very little radiation at these longer wavelengths, it traps this thermal radiation.

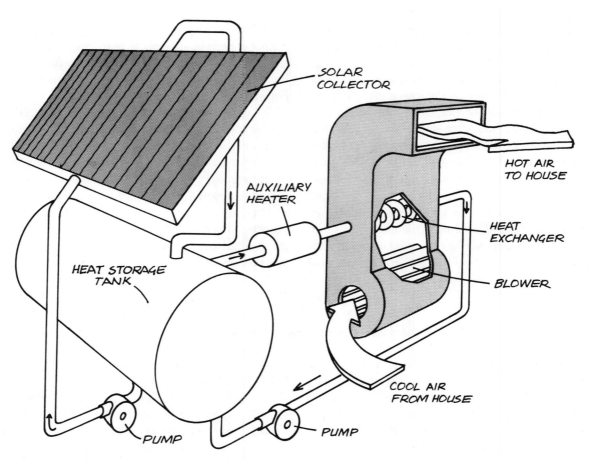

SOLAR COLLECTOR

AUXILIARY HEATER

HOT AIR TO HOUSE

HEAT EXCHANGER

BLOWER

HEAT STORAGE TANK

COOL AIR FROM HOUSE

PUMP

PUMP

A typical active system for solar heating.

SOLAR HEATING METHODS

The great variety of methods used to trap solar radiation for home heating can be grouped into two broad categories—passive and active. In *passive* systems, the sun's energy is collected, stored, and transmitted without the use of electrical or mechanical energy. Passive systems can be further subdivided into direct gain and indirect gain systems. *Direct gain* systems are the simplest way to solar heat. They require at most a rearrangement of standard construction practices. Almost all solar homes employ some direct gain, unless poor orientation or unsightly views prohibit it.

Indirect gain systems collect the sun's energy before it enters the home. Then they either di-

rect the heat into the building to be stored there, or use ingenious adaptations of the natural thermal properties of materials to store and distribute the heat. The energy flows to rooms without the help of complex ducts, piping, or pumps. Such systems are often an integral part of the home itself. Although they may call for nonstandard building practices, they can be simple and effective.

Active systems for solar heating generally use rooftop solar collectors and separate heat storage devices, although if small enough, they too can use the mass of the house itself for storage. Heat moves from the collectors to storage or to interior spaces through pipes or ducts. Pumps or fans circulate a fluid through the collector and back to the house or to an insulated heat storage container. In the second case, if the house needs heat, the fluid from the central heating system is warmed by the stored heat and circulated through the rooms. Such heating systems are called *active* because they rely on mechanical and electrical power to move the heat.

Most active solar heating systems use an array of *flat-plate collectors* to gather solar energy. These collectors have one or more glass or plastic cover plates with a black absorber beneath them. The cover plates reduce the loss of energy through the front, and insulation behind the absorber reduces the heat loss through the back. Heat from the absorber is conducted to a transfer fluid, either a gas (usually air) or a liquid (water or antifreeze), which flows in contact with it and carries off the heat.

In *concentrating collectors,* reflective surfaces concentrate the sun's rays onto a very small area—often an evacuated tube. This solar energy is then absorbed by a black surface and converted to heat that is carried off by a fluid. Concentrating collectors can produce very high temperatures, and some require mechanical devices to track the sun across the sky. They are most often seen in large scale applications, such as industrial heating or generation of electricity.

Depending on the climate, the house, and the solar heating system design, 50 to 90 percent of a house's heating needs can be readily supplied by the sun. However, solar heating systems almost always require a backup, or auxiliary heating system. Rarely is it economical to build a heat storage unit with the capacity to carry a house through long periods of cold and cloudy weather.

OTHER SOLAR APPLICATIONS

Two other uses of sunlight have a strong place in the market: systems for heating domestic hot water and attached greenhouse solariums called sunspaces. A third application, photovoltaics, is still struggling to achieve a cost-benefit ratio that will attract major attention, but it has long-term promise.

Solar heating of domestic hot water (DHW) is a smaller scale application of the same concepts and techniques used for home heating. It can have a lower first cost and can fit in easily with existing conventional water heating systems.

Sunspaces are a modern version of traditional sunporches or attached greenhouses, designed to serve many purposes. Depending on the particular design combination, sunspaces can be attractive living spaces, economical sources of auxiliary heat, a place for growing plants, or a combination of all three.

In photovoltaics, a way of getting electricity directly from the sun, solar cells use the semiconducting properties of materials such as silicon to convert sunlight to electricity. Photovoltaics has enormous potential. At present, however, only in remote areas can solar cells compete on overall cost with other methods of generating electricity.

Using sunlight for heat and energy goes back a long way in human history. But the last forty years have seen the most dramatic progress in developing solar technology. The purpose of this book is to present the principles of solar design, so that you can understand how and why these principles can be applied to using the free and abundant energy of the sun.

I
Fundamentals

Is it not by the vibrations given to it by the sun that light appears to us; and may it not be that every one of the infinitely small vibrations, striking common matter with a certain force, enters its substance, is held there by attraction and augmented by successive vibrations, till the matter has received as much as their force can drive into it?

Is it not thus that the surface of this globe is heated by such repeated vibrations in the day, and cooled by the escape of the heat when those vibrations are discontinued in the night?

Benjamin Franklin,
Loose Thoughts on a Universal Fluid

Before you design and build a solar home, you need to become familiar with your surroundings. You need to know the position of the sun in order to orient a house or collector to receive its warm rays. To gauge the solar heat flows into a house you must calculate the solar radiation hitting the walls, windows, roofs and collector surfaces. You also need to calculate the heat escaping from a house in order to select the best methods to slow it down. Only when you have grasped the fundamentals can you take advantage of these natural energy flows.

First you need to understand some of the language others use to describe and measure energy. Become familiar with climatic data and the properties of common building materials. The aim of this is to aquaint you with these and other essentials that will help you use the abundance of solar energy falling all around you. Some of this may seem tedious, but it is all very important to good solar home design.

1
Solar Phenomena

After centuries of observation, ancient astronomers could accurately predict the sun's motion across the sky. Stonehenge was probably a gigantic "computer" that recorded the movements of the sun and moon in stone. From their earthbound viewpoint, early peoples reckoned that the sun gave them night and day by moving in a path around the earth. But today, thanks to the work of the sixteenth-century Polish astronomer Copernicus, we know that the earth travels in an orbit around the sun and that the rotation of the earth, not the motion of the sun, gives us the cycles of night and day.

The earth actually follows an elliptical (egg-shaped) path around the sun. As it travels this orbit, its distance from the sun changes slightly—it is closest in winter and most distant in summer. The amount of solar radiation striking the earth's atmosphere is consequently most intense in winter. Then why are winters so dreadfully cold?

This seeming paradox is readily explained. The earth's axis is tilted relative to the plane of its orbit, as shown in the first diagram. The north pole is tilted *toward* the sun in summer and *away from* the sun in winter. This angle is called the *declination* angle. From our viewpoint here on earth, this tilt means that the sun is higher in the sky in summer, and lower in winter. Consequently, the sun's rays have a

greater distance to travel through the atmosphere in winter, and they strike the earth's surface at a more glancing angle. The amount of solar radiation eventually striking a horizontal surface is less during the winter, and the weather is colder.

This tilt of the earth's axis results in the seasons of the year. If the axis were perpendicular to the orbital plane, there would be no noticeable change of seasons. Each day the sun would follow the same path across the sky, and the weather would be uniformly dull. Likewise, if the earth did not rotate on its axis, the sun would creep slowly across the sky, and a single day would last a whole year. The diurnal (daily) and seasonal cycles that we take for granted are a direct result of this rotation of the earth about a tilted axis.

SOLAR POSITION

Most people have probably noticed that the sun is higher in the sky in summer than in winter. Some also realize that it rises south of due east in winter and north of due east in summer. Each day the sun travels in a circular path across the sky, reaching its highest point at noon. As winter proceeds into spring and summer, this circular path moves higher in the sky. The sun rises earlier in the day and sets later.

9

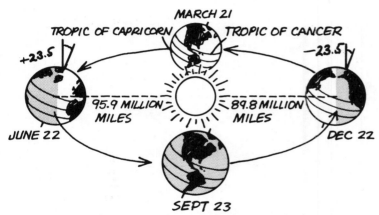

The earth's elliptical path around the sun. The tilt of the earth's axis results in the seasons of the year. The declination angles on June 22 and Dec. 22 are +23.5 and −23.5, respectively. The declination angles on Mar. 21 and Sept. 23 are both 0.

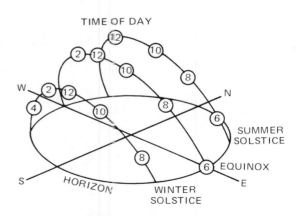

The sun's daily path across the sky. The sun is higher in the sky in summer than in winter due to the tilt of the earth's axis.

The actual position of the sun in the sky depends upon the latitude of the observer. At noon on March 21 and September 23, the vernal and autumnal *equinoxes*, the sun is directly overhead at the equator. At 40°N latitude, however, its angle above the horizon is 50° (90 − 40°). By noon on June 22, the *summer solstice* in the Northern Hemisphere, the sun is directly overhead at the Tropic of Cancer, 23.5°N latitude. Its angle above the horizon at 40°N is 73.5° (90° + 23.5° − 40°), the highest it gets at this latitude. At noon on December 22, the sun is directly overhead at the Tropic of Capricorn, and its angle above the horizon at 40°N latitude is only 26.5° (90° − 23.5° − 40°).

A more exact description of the sun's position is needed for most solar applications. In the language of trigonometry, this position is expressed by the values of two angles—the solar altitude and the solar azimuth. The solar *altitude* (represented by the Greek letter theta θ) is measured up from the horizon to the sun, while the solar *azimuth* (the Greek letter phi φ) is the angular deviation from true south.

These angles need not be excessively mysterious—you can make a rough measurement of them with your own body. Stand facing the sun with one hand pointing toward it and the other pointing due south. Now drop the first hand so that it points to the horizon directly below the sun. The angle that your arm drops

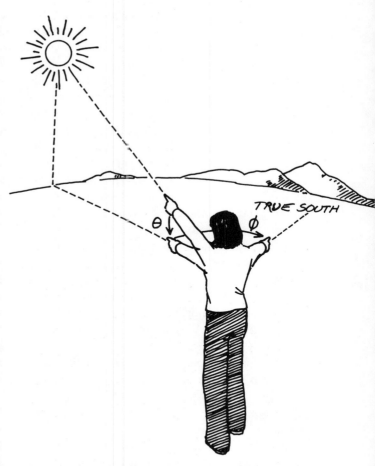

Measuring the sun's position. The solar altitude (theta θ) is the angle between the sun and the horizon, and the azimuth (phi φ) is measured from true south.

10

SOLAR POSITIONS FOR 40°N LATITUDE

AM	PM	Jan 21	Feb 21	Mar 21	Apr 21	May 21	Jun 21	Jul 21	Aug 21	Sep 21	Oct 21	Nov 21	Dec 21
5	7					1.9 / 114.7	4.2 / 117.3	2.3 / 115.2					
6	6					7.4 / 98.9	12.7 / 105.6	14.8 / 108.4	13.1 / 106.1	7.9 / 99.5			
7	5		4.3 / 72.1	11.4 / 80.2	18.9 / 89.5	24.0 / 96.6	26.0 / 99.7	24.3 / 97.2	19.3 / 90.0	11.4 / 80.2	4.5 / 72.3		
8	4	8.1 / 55.3	14.8 / 61.6	22.5 / 69.6	30.3 / 79.3	35.4 / 87.2	37.4 / 90.7	35.8 / 87.8	30.7 / 79.9	22.5 / 69.6	15.0 / 61.9	8.2 / 55.4	5.5 / 53.0
9	3	16.8 / 44.0	24.3 / 49.7	32.8 / 57.3	41.3 / 67.2	46.8 / 76.0	48.8 / 80.2	47.2 / 76.7	41.8 / 67.9	32.8 / 57.3	24.5 / 49.8	17.0 / 44.1	14.0 / 41.9
10	2	23.8 / 30.9	32.1 / 35.4	41.6 / 41.9	51.2 / 51.4	57.5 / 60.9	59.8 / 65.8	57.9 / 61.7	51.7 / 52.1	41.6 / 41.9	32.4 / 35.6	24.0 / 31.0	20.7 / 29.4
11	1	28.4 / 16.0	37.3 / 18.6	47.7 / 22.6	58.7 / 29.2	66.2 / 37.1	69.2 / 41.9	66.7 / 37.9	59.3 / 29.7	47.7 / 22.6	37.6 / 18.7	28.6 / 16.1	25.0 / 15.2
12		30.0 / 0.0	39.2 / 0.0	50.0 / 0.0	61.6 / 0.0	70.0 / 0.0	73.5 / 0.0	70.6 / 0.0	62.3 / 0.0	50.0 / 0.0	39.5 / 0.0	30.2 / 0.0	26.6 / 0.0

Notes: Top number in each group is altitude angle, measured from the horizon. Second number is azimuth angle, measured from true south. Angles given in degrees, and solar times used.

is the solar altitude (θ) and the angle between your arms in the final position is the solar azimuth (ϕ). Much better accuracy can be obtained with better instruments, but the measurement process is essentially the same.

The solar altitude and azimuth can be calculated for any day, time, and latitude. For 40°N latitude (Philadelphia, for example) the values of θ and ϕ are given at each hour for the 21st day of each month in the accompanying table. Note that ϕ is always zero at solar noon and the θ varies from 26.6° at noon on December 21 to 73.5° at noon on June 21. You can find similar data for latitudes 24°N, 32°N, 48°N, 56°N, and 64°N in the table titled "Clear Day Insolation Data" in the appendix. This appendix also shows you how to calculate these angles directly for any day, time, and latitude.

Why do you need to know these solar positions? A knowledge of the sun's position helps you determine the orientation of a house and placement of windows to collect the most winter sunlight. This knowledge is also helpful in positioning shading devices and vegetation to block the summer sun. Often the available solar radiation data only applies to horizontal or south-facing surfaces, and exact solar positions are needed to convert these data into values that are valid for other surfaces.

INSOLATION

Arriving at a quantitative description of the solar radiation striking a surface, or the *insolation* (not to be confused with insulation), is a difficult task. Most of this difficulty arises from

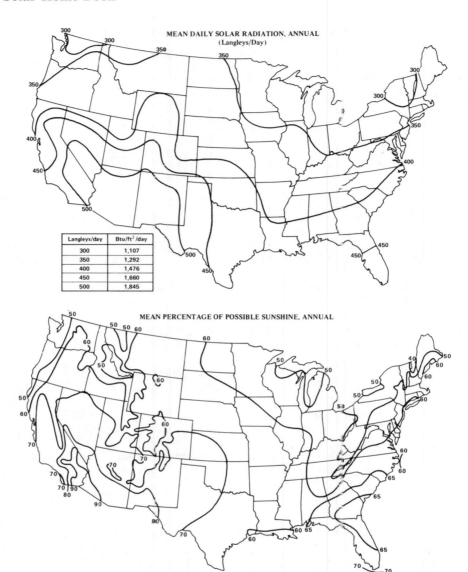

Langleys/day	Btu/ft² /day
300	1,107
350	1,292
400	1,476
450	1,660
500	1,845

the many variables that affect the amount of solar radiation striking a particular spot. Length of day, cloudiness, humidity, elevation above sea level, and surrounding obstacles all affect the insolation. Compounding this difficulty is the fact that the total solar radiation striking a surface is the sum of three contributions: the *direct radiation* from the sun, the *diffuse radiation* from the entire sky, and the *reflected radiation* from surrounding terrain, buildings, and vegetation. Fortunately, however, we do not need exact insolation data for most low-temperature applications of solar energy.

Although insolation data has been recorded at about 80 weather stations across the country, much of it is inaccurate and incomplete. The information is usually provided in units of *langleys* striking a horizontal surface over a period of time, usually a day. A langley is one calorie of radiant energy per square centimeter, and one langley is equivalent to 3.69 Btu per square foot, the more familiar English measure. An example of the information available is the map of ''Mean Daily Solar Radiation, Annual'' presented here. You can find monthly maps of the mean daily solar radiation in the appendix. These

Diffuse and Reflected Radiation

The total solar radiation striking a surface is the sum of three components: the direct solar radiation (I_D), the diffuse sky radiation (I_d), and the radiation reflected from surroundings (I_r). The direct component consists of rays coming straight from the sun—casting strong shadows on a clear day. If all our days were clear, we could simply use the Clear Day Insolation Data, add a small percentage for ground reflection, and have a very good estimate of the total insolation on our walls, roofs, and collectors. But all of us can't live in Phoenix or Albuquerque, so we must learn to deal with cloudy weather.

As it passes through the atmosphere, sunlight is scattered by air molecules, dust, clouds, ozone, and water vapor. Coming uniformly from the entire sky, this scattered radiation makes the sky blue on clear days and grey on hazy days. Although this diffuse radiation amounts to between 10 and 100 percent of the radiation reaching the earth's surface, little is known about its strength and variability.

The Clear Day Insolation Data aren't much help on a cloudy day. But frequently we only need to know the average daily insolation over a period of a month. In such a case we can use the monthly maps of the percent of possible sunshine to help us estimate this average. If P is the percentage of possible sunshine for the month and location in question, then we compute a factor F according to

$$F = 0.30 + 0.65(P/100)$$

The numbers 0.30 and 0.65 are coefficients that actually vary with climate, location, and surface orientation. But their variation is not too severe, and we can use these average values for estimating average daily insolation. If I_o is the Clear Day Insolation (whole day total) on a plane surface, then we compute the average daily insolation (I_a) according to

$$I_a = F(I_o)$$

These formulas estimate the diffuse radiation that still strikes the surface on cloudy and partly cloudy days. Even in a completely cloudy month (P = 0), we would still be receiving 30 percent (F = 0.30) of the clear day insolation, according to these equations. This is perhaps a bit high, but the coefficients have been selected to produce accurate results under normal conditions, not blackouts.

For example, calculate the average daily insolation striking a horizontal roof in Philadelphia during the months of June and January. Using the first equation and P = 65 (June) and 49 (January) from before, we get for June:

$$F = 0.30 + 0.65(65/100) = 0.72$$

For January:

$$F = 0.30 + 0.65(49/100) = 0.62$$

Therefore, the average daily insolation is, for June:

$$I_a = 0.72(2618) = 1907 \ Btu/ft^2$$

For January:

$$I_a = 0.62(948) = 588 \ Btu/ft^2$$

These numbers may be compared with the 1721 Btu/ft^2 and 464 Btu/ft^2 calculated earlier. If we include diffuse radiation during cloudy weather, our results are 10 to 20 percent higher than before.

The diffuse and reflected radiation striking a surface also depend upon the orientation of the surface. Under the same sky conditions, a horizontal roof (which "sees" the entire sky) receives about twice the diffuse radiation hitting a vertical wall (which "sees" only one half the sky). Tilted surfaces receive some average of these two. Ground reflection depends a lot upon the shape and texture of the surroundings and the altitude of the sun. Snow reflects much more sunlight than green grass, and more reflection occurs when the sun is lower in the sky. During the winter, as much as 30 percent of the horizontal clear day insolation may be reflected up onto the surface of a south facing wall. But a roof receives no reflected radiation in any season, because it faces the sky, not the ground.

13

CLEAR DAY INSOLATION FOR 40°N LATITUDE

TOTAL INSOLATION, Btu/ft^2

21st Day	Normal Surface	Horizontal Surface	South facing surface tilt angle				
			30°	40°	50°	60°	90°
January	2182	948	1660	1810	1906	1944	1726
February	2640	1414	2060	2162	2202	2176	1730
March	2916	1852	2308	2330	2284	2174	1484
April	3092	2274	2412	2320	2168	1956	1022
May	3160	2552	2442	2264	2040	1760	724
June	3180	2648	2434	2224	1974	1670	610
July	3062	2534	2409	2230	2006	1728	702
August	2916	2244	2354	2258	2104	1894	978
September	2708	1788	2210	2228	2182	2074	1416
October	2454	1348	1962	2060	2098	2074	1654
November	2128	942	1636	1778	1870	1908	1686
December	1978	782	1480	1634	1740	1796	1646

data apply only to horizontal surfaces, and can be misleading. Complicated trigonometric conversions, which involve assumptions about the ratio of direct to diffuse radiation, are necessary to apply these data to vertical or tilted surfaces. The trigonometric conversions are also discussed the the appendix.

The weather bureau also provides information about the percentage of possible sunshine, defined as the percentage of time the sun "casts a shadow." An example of these data is the map shown here titled "Mean Percentage of Possible Sunshine, Annual." In the appendix you will find monthly maps that are more useful for calculations of insolation. By themselves, these maps tell us little about the amount of solar radiation falling on a surface, but when coupled with the "Clear Day Insolation Data," they make a powerful design tool.

Clear Day Insolation tables, prepared by the American Society of Heating, Refrigerating, and Air-Conditioning Engineers (ASHRAE), provide hourly and daily insolation (and solar positions) for a variety of latitudes. Tables for 24°N, 32°N, 40°N, 48°N, and 36°N latitude are reprinted in the appendix. The values of the daily insolation from the 40°N latitude table are included here as an example. These tables list the *average clear day insolation* on horizontal and normal (perpendicular to the sun) surfaces, and on five south-facing surfaces tilted at different angles (including vertical). The insolation figures quoted include a diffuse contribution for an "average" clear sky, but do not include any contribution for reflections from the surrounding terrain.

Hourly and daily insolation data are given in the appendix for the 21st day of each month. You can readily interpolate between these numbers to get values of the insolation for other days, times, latitudes, and south-facing orientations. Trigonometric conversions of these data to other surface orientations are explained there.

When multiplied by the appropriate "percentage of possible sunshine," these data provide an estimate of the hourly and daily insolation on a variety of surface orientations. You will note, for example, that the total clear day insolation on a vertical south-facing wall in Philadelphia (40°N) is 610 Btu/ft^2 on June 21 and

1726 Btu/ft^2 on January 21—almost three times greater! Multiplied by the percentage of possible sunshine for this locale (about 65% in June and 49% in January), the total insolation becomes 396 Btu/ft^2 in June and 846 Btu/ft^2 in January, or still a factor of two greater. On the other hand, the clear day insolation on a horizontal roof is 2648 Btu/ft^2 in June and only 948 Btu/ft^2 in January, or almost a factor of four smaller. Clearly, the roof is taking the heat in summer and the south walls are getting it in winter.

LIMITATIONS OF INSOLATION DATA

You must be careful to note the limitations of the Clear Day Insolation table. These data are based upon "average" clear day conditions, but "average" can vary with locale. Many locations are 10 percent clearer, such as deserts and mountains, and others, such as industrial and humid areas are not as clear as the "average." Reflected sunlight from vegetation and ground cover is not included in the values given in the tables. Another 15 to 30 percent more sunlight may be reflected onto a surface than the amount listed. In the winter, even more radiation will be reflected onto south-facing walls because the sun is lower in the sky and snow may be covering the ground.

Other difficulties arise from the subjective evaluations of "percentage of possible sunshine." In the method of calculating average insolation described above, an assumption was made that the sun is shining full blast during the "sunshine" period and not at all during other times. In reality, up to 20 percent of the clear day insolation may still be hitting the surface during periods of total cloudiness. During hazy periods when the sun still casts a shadow, only 50 percent of the clear day insolation may be striking the surface. More accurate calculations, in which the diffuse and direct components of solar radiation are treated separately, are provided in the appendix.

Another problem is the variability of weather conditions with location and time of day. The weather maps provide only area-wide averages of the percent of possible sunshine. The actual value in your exact building location could be very different from your county average. On the other hand, the cloudiness in some areas, particularly coastal areas, can occur at specific times of the day, rather than being distributed at random over the entire day. There may be a morning fog when the sun is low on the horizon, and a clear sky from mid-morning on, but this would be recorded as 75 percent of possible sunshine, while 90 percent of the total clear day insolation was actually recorded that day.

You may need more detailed information than is available from national weather maps. Occasionally, friendlier-than-usual personnel will assist you at the local weather station, but you will almost always be referred to the National Weather Records Center in Asheville, North Carolina. This center collects, stores, and distributes weather data from around the country, and makes it available in many forms. You should first obtain their "Selective Guide to Climate Data Sources," to give you an overview of the types of data available. You may obtain a copy from the Superintendent of Documents there.

2
Heat Flow Calculations

Heat energy is simply the motion of the atoms and molecules in a substance—their twirling, vibrating, and banging against each other. It is this motion that brings different atoms and molecules together in our bodily fluids, allowing the chemical reactions that sustain us. This is why our bodies need warmth. Seventeenth-century natural philosphers thought heat was a fluid—"phlogiston" they called it—that was released by fire and flowed from hot bodies to cold. They were correct about this last observation, for heat always flows from warm areas to colder ones.

The rate of heat flow is proportional to the temperature difference between the source of the heat and the object or space to which it is flowing. Heat flows out of a house at a faster rate on a cold day than on a mild one. If there is no internal source of heat, such as a furnace or wood stove, the temperature inside the house approaches that of the outdoor air. Heat always flows in a direction that will equalize temperatures.

While the rate of heat flow is proportional to the temperature difference, the quantity of heat actually flowing depends on how much resistance there is to the flow. Since we can do little about the temperature difference between inside and outside, most of our effort goes into increasing a building's resistance to heat flow.

The actual mechanisms of heat flow are numerous, and so are the methods of resisting them. Therefore, we will review briefly the three basic methods of heat flow—conduction, convection and radiation.

As children, we all learned about heat conduction intuitively by touching the handle of a hot skillet. When an iron skillet sits on a hot stove for a while, heat from the burner flows through the metal of the skillet to the handle. But the rate of flow to the handle of an iron skillet is much slower than if the skillet were made of copper. The heat flow through copper is quicker because it has a greater conductance (less resistance to heat flow) than cast iron. It also takes less heat to warm copper than iron, and therefore less time to heat the metal between the burner and the handle. These principles are basic to the concept of conduction heat flow.

Convection is heat flow through the movement of fluids—liquids or gases. In a kettle of water on a stove, the heated water at the bottom rises and mixes with the cooler water above, spreading the heat and warming the entire volume of water far more quickly than could have been done by heat conduction alone. A house with a warm air furnace is heated in much the same way. Air is heated in the firebox and rises up to the living spaces. Since the house air is cooler than the hot furnace air, the heat is transferred from the hot furnace air to the cooler room air and then to the surfaces in the rooms.

Heated fluids can move by natural convection or forced convection. As a fluid is warmed, it expands and becomes less dense, making it buoyant in the surrounding cooler fluid. it rises and the cooler fluid that flows in to replace it is heated in turn. The warmed fluid moves to a cooler place where its heat is absorbed. Thus the fluid cools down, becomes heavier and sinks. This movement is known as *natural convection* or *thermosiphoning*. When we want more control over the heat flow, we use a pump or a blower to move the heated fluid. This is called *forced convection*.

Note that convection works hand-in-hand with conduction. Heat from a warm surface is conducted to the adjacent fluid before it is carried away by convection, and heat is also conducted from a warm fluid to a cool surface nearby. The greater the temperature difference between the warm and cool surfaces, the greater the heat flow between them.

Thermal radiation is the flow of heat energy through an open space by electromagnetic waves. This flow occurs even in the absence of any material in that space—just as sunlight can leap across interplanetary voids. Objects that stop the flow of light also stop thermal radiation, which is primarily invisible longwave radiation. Warmer objects constantly radiate their thermal energy to cooler objects (as long as they can "see" each other) at a rate proportional to their temperature difference.

We experience radiative heat flow to our bodies when we stand in front of a fireplace or hot stove. The same transfer mechanism, although more subtle and difficult to perceive, is what makes us feel cold while sitting next to a window on a winter night. Our warm bodies are radiating energy to the cold window surface, and we are chilled.

Of the three basic kinds of heat loss, radiation is the most difficult to calculate at the scale of a house. Calculation of convection heat loss through open doors or cracks and around window frames is educated guesswork. Conduction

heat loss through the exterior skin of the house (roofs, walls, and floors) is perhaps the easiest to estimate. Fortunately, this is the thief that can pilfer the most heat from our homes.

CONDUCTION HEAT LOSS

The ability of a material to permit the flow of heat is called its thermal conductivity or conductance. The *conductance* (C) of a slab of material is the quantity of heat that will pass through one square foot of that slab per hour with a 1°F temperature difference maintained between its two surfaces. Conductance is measured in units of Btu per hour per square foot per degree Fahrenheit, or Btu/(hr ft² °F).

The total conductance of a slab of material decreases as its thickness increases. While 10 Btu per hour may flow through a 1-inch slab of polystyrene, only 5 Btu per hour will flow through a 2-inch slab under the same conditions.

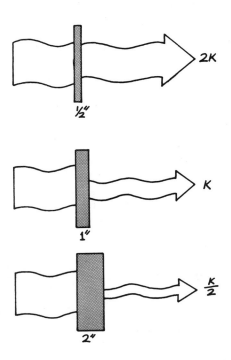

The thicker a slab, the less heat it conducts.

The New Solar Home Book

The opposite of conductance is *resistance*, the tendency of a material to retard the flow of heat. All materials have some resistance to heat flow—those with high resistance we call insulation. The *resistance* (R) of a slab of material is the inverse of its conductance, R = (1/C). The higher the R-value of a material, the better its insulating properties. R-values are expressed in (hr ft^2 °F)/Btu. In the table you can find R-values for a few common building materials. More detailed lists are provided in the appendix under "Insulating Value of Materials."

A related quantity, the overall *coefficient of heat transmission* (U), is a measure of how well a wall, roof, or floor conducts heat. The lower the U-value of a wall, the higher its insulating ability. Numerically, U is the rate of heat loss in Btu per hour through a square foot of surface with a 1 degree (°F) temperature difference between the inside and outside air. Similar to conductance, U is expressed in units of Btu/(hr ft^2 °F). To find the conduction heat loss (ΔH_{con}), through an entire wall, we multiply its U value by the number of hours (h), the wall area (A), and the temperature difference (ΔT), between the inside and outside air:

$$\Delta H_{con} = (U)(h)(A)(\Delta T)$$

SAMPLE CALCULATIONS OF U-VALUES

Wall Construction Components	R-values Uninsulated	Insulated
Outside air film, 15 mph wind	0.17	0.17
0.75" beveled wood siding, lapped	0.81	0.81
0.50" plywood sheathing	0.62	0.62
3.5" air space	1.01	-
3.5" mineral fiber batt	-	11.00
0.5" gypsum board	0.45	0.45
Inside air film	0.68	0.68
TOTALS (R_t)	3.74	13.73
U-Values (U = 1/R_t)	0.27	0.07

RESISTANCES OF COMMON BUILDING MATERIALS

Material	Thickness (inches)	R-Value (ft^2 °F hr)/Btu
Hardwood (oak)	1.0	0.91
Softwood (pine)	1.0	1.25
Gypsum board	0.5	0.45
Wood shingles	lapped	0.87
Wood bevel siding	lapped	0.81
Brick, common	4.0	0.80
Concrete (sand and gravel)	8.0	0.64
Concrete blocks (filled cores)	8.0	1.93
Gypsum fiber concrete	8.0	4.80
Mineral fiber (batt)	3.5	11.00
Mineral fiber (batt)	6.0	19.00
Molded polystyrene beads	1.0	3.85
Fiberglass board	1.0	4.35
Extruded polystyrene	1.0	5.00
Cellular polyurethane	1.0	6.25
Polyisocyanurate	1.0	7.04
Phenolic foam	1.0	8.33
Loose fill insulation:		
Cellulose fiber	1.0	3.13-3.70
Mineral wool	1.0	2.93
Sawdust	1.0	2.22
Flat glass	0.125	0.91
Insulating glass (0.25" space)		1.69

SOURCE: *ASHRAE Handbook, 1985 Fundamentals.*

To find the heat loss through a 50 sq ft wall with a U-value of 0.12 over an 8–hour time span, when the inside temperature is 65°F and the outside temperature is 40°F, multiply:

$$\Delta H_{con} = (0.12)(8)(50)(65 - 40) = 1200 \text{ Btu}$$

If the inside temperature is 70°F instead of 65°F, then the heat loss is 1440 Btu over the same time span.

The U-value includes the thermal effects of all the materials in a wall, roof, or floor—including air gaps inside, and air films on the inner and outer surfaces. It can be computed from the conductances or resistances of all these separate components. The total resistance R_t is the sum of the individual resistances of these components. As U is the conductance of the entire building section, it is the inverse of R_t, or

18

$$U = (1/R_t) = 1/(R_1 + R_2 + R_3 + \ldots + R_n)$$

Thus, computation of U involves adding up all the R-values, including R-values of inside and outside air films, any air gap greater than three quarters of an inch, and all building materials.

As an example, the U-values of two typical walls, one insulated and the other uninsulated, are calculated here. Note that the uninsulated wall conducts heat almost four times more rapidly than the insulated wall.

This is a simplified version of the heat flows. Heat will pass more quickly through the framing of the wall than through the insulation. If the total R-value through the framing section of the wall is 7.1, and the framing takes up 20 percent of the wall cavity, then the weighted R-value of the insulated wall is:

$$R_w = 0.20(7.1) + 0.80(13.73) = 12.4$$

The weighted R-value of the uninsulated wall is:

$$R_w = 0.20(7.1) + 0.80(3.74) = 4.4$$

Notice that the weighted R-value of the insulated wall is now less than three times better than the uninsulated wall.

Once you have calculated the U-values of all exterior surfaces (windows, walls, roofs, and floors) in a house, you can begin calculating the total conduction heat loss. One important quantity is the hourly heat loss of the house at outside temperatures close to the lowest expected. These extreme temperatures are called *design temperatures*. A list of the recommended design temperatures for a number of U.S. cities is provided here; those for many other locations in the United States are provided in the appendix under "Degree Days and Design Temperatures."

The following approach is used to find the Btu per hour your heating system will have to supply in order to keep your house warm under all but the most extreme conditions. Subtract the design temperature from the normal inside temperature to find the temperature difference (ΔT). Next, determine the total area (A) of each type of exterior building surface and multiply it by the temperature difference and the appropriate U-value (U_s), to get the total conduction heat loss (ΔH_s) of that surface per hour:

$$\Delta H_s = U_s(A_s)(\Delta T)$$

The total conduction heat loss of the house is merely the sum of the conduction heat losses through all these building surfaces. For example, the conduction heat loss of the 50-square foot insulated wall with a U-value of 0.07 under design temperature conditions ($-2°F$) in Denver, Colorado, is

$$\Delta H_s = 0.07(50)[70 - (-2)] = 252 \text{ Btu/hr.}$$

To compute the total conduction heat loss for a single heating season, you must first grasp the concept of degree days. They are somewhat analogous to man-days of work. If a man works one day, the amount of work he does is often called a man-day. Similarly, if the outdoor temperature is one degree below the indoor temperature of a building for one day, we say one *degree day* (D) has accumulated.

Standard practice uses an indoor temperature of 65°F as the base from which to calculate degree days, because most buildings do not require heat until the outdoor air temperature falls between 60°F and 65°F. If the outdoor temperature is 40°F for one day, then $65 - 40 = 25$ degree days result. If the outdoor temperature is 60°F for five days, then $5(65 - 60) = 25$ degree days again result. (When we refer to degree days here, we mean degrees Farenheit (°F), unless otherwise noted.)

The Weather Service publishes degree day information in special maps and tables. Maps showing the monthly and yearly total degree days are available in the *Climatic Atlas*. Tables of degree days, both annual and monthly, are provided for many cities in the appendix under "Degree Days and Design Temperatures." Your local oil dealer or propane distributor should also know the number of degree days for your town.

To compute the total conduction heat loss during the heating season, you first multiply the total degree days for your locality by 24 (hours

The New Solar Home Book

DEGREE DAYS AND DESIGN TEMPERATURES
(HEATING SEASON)

State	City	Design Temperature (°F)	Degree Days (°F days)	State	City	Design Temperature (°F)	Degree Days (°F days)
Alabama	Birmingham	19	2,600	Nevada	Reno	2	6,300
Alaska	Anchorage	-25	10,900	New Hampshire	Concord	-11	7,400
Arizona	Phoenix	31	1,800	New Mexico	Albuquerque	14	4,300
Arkansas	Little Rock	19	3,200	New York	Buffalo	3	7,100
California	Los Angeles	41	2,100	New York	New York	11	4,900
California	San Francisco	42	3,000	North Carolina	Raleigh	16	3,400
Colorado	Denver	-2	6,300	North Dakota	Bismarck	-24	8,900
Connecticut	Hartford	1	6,200	Ohio	Columbus	2	5,700
Florida	Tampa	36	700	Oklahoma	Tulsa	12	3,900
Georgia	Atlanta	18	3,000	Oregon	Portland	21	4,600
Idaho	Boise	4	6,200	Pennsylvania	Philadelphia	11	5,100
Illinois	Chicago	-4	6,600	Pennsylvania	Pittsburgh	5	6,000
Indiana	Indianapolis	0	5,700	Rhode Island	Providence	6	6,000
Iowa	Des Moines	-7	6,600	South Carolina	Charleston	26	1,800
Kansas	Wichita	5	4,600	South Dakota	Sioux Falls	-14	7,800
Kentucky	Louisville	8	4,700	Tennessee	Chattanooga	15	3,300
Louisiana	New Orleans	32	1,400	Texas	Dallas	19	2,400
Maryland	Baltimore	12	4,700	Texas	San Antonio	25	1,500
Massachusetts	Boston	6	5,600	Utah	Salt Lake City	5	6,100
Michigan	Detroit	4	6,200	Vermont	Burlington	-12	8,300
Minnesota	Minneapolis	-14	8,400	Virginia	Richmond	14	3,900
Mississippi	Jackson	21	2,200	Washington	Seattle	28	4,400
Missouri	St. Louis	4	4,900	West Virginia	Charleston	9	4,500
Montana	Helena	-17	8,200	Wisconsin	Madison	-9	7,900
Nebraska	Lincoln	-4	5,900	Wyoming	Cheyenne	-6	7,400

per day) to get the total *degree hours* during that time span. Now your calculation proceeds as in the earlier example: you multiply the area of each section (A_s) by its U-value (U_s) and the number of degree hours (24D) to get the seasonal heat loss through that section:

$$\text{Seasonal } \Delta H_s = A_s (U_s)(24)(D)$$

The seasonal conduction heat loss from the entire house is the sum of seasonal heat losses through all the building surfaces. A short cut is to multiply the U-value of each section times the area of each section to get the "UA" for that section. Add together all the UA's and then multiply by 24D to get the total seasonal conductive heat loss:

$$\text{Seasonal } \Delta H = (UA_1 + UA_2 + UA_3 \ldots + UA_n)(24)(D)$$

But to get the total seasonal heat loss, you must include the convection heat losses described in the next section.

CONVECTION HEAT LOSS

There are three modes of convection which influence the heat loss from a building. The first two have already been included in the calculation of conduction heat losses through the building skin. They are the convection heat flow across air gaps in the wall and heat flow to or from the walls through the surrounding air. These two effects have been included in the calculation of U-values by assigning insulating values to air gaps or air films. The third mode of convection heat flow is *air infiltration* through openings in walls (such as doors and windows) and through cracks around doors and windows. In a typical house, heat loss by air infiltration is often comparable to heat loss by conduction.

The first mode of convection heat loss occurs within the walls and between the layers of glass in the skin of the building. Wherever there is an air gap, and whenever there is a temperature difference between the opposing surfaces of that gap, natural air convection results in a heat flow across that gap. This process is not very efficient, so air gaps are considered to have some insulating value. For the insulating value to be significant, the width of the air gap must be greater than 3/4 inch. However, a quick glance at the insulating values of air gaps in the appendix reveals that further increases in the width don't produce significant increases in insulation. Wider air gaps allow freer circulation of the air in the space, offsetting the potentially greater insulating value of the thicker air blanket.

Most common forms of insulation do their job simply by trapping air in tiny spaces to prevent air circulation in the space they occupy. Fiberglass blanket insulation, rigid board insulation, cotton, feathers, crumpled newspaper, and even popcorn make good insulators because they create tiny air pockets to slow down the convection flow of heat.

Conduction heat flow through the exterior skin of a house works together with air movements within the rooms and winds across the exterior surface to siphon off even more heat.

Interior surfaces of uninsulated perimeter walls are cooler than room air. They cool the air film right next to the wall. This cooled air sinks down and runs across the floor, while warmer air at the top of the room flows in to take its place, accelerating the cooling of the entire room. The inside surface of a well-insulated wall will have about the same temperature as the room air. But the inside surface of a window will be much colder, and the air movement and cooling effects are severe. Heating units or warm air registers have traditionally been placed beneath windows in an effort to eliminate the cold draft coming down from the glass surfaces. While this practice improves the comfort of the living areas, it substantially increases the heat losses to the outdoors. With the advent of new, higher R-value glazing materials, better insulated walls, and lower infiltration rates, this location isn't as important in energy-conserving home.

Though not very large, the insulating value of the air films on either side of a wall or roof do make a contribution to the overall U-value. The air films on horizontal surfaces provide more insulation than those on vertical surfaces. (Convection air flow, which reduces the effective thickness of the still air insulating film, is greater down a vertical wall than across a horizontal surface.) Similarly, the air film on the outside surface is reduced by wind blowing across the surface. The higher the wind speed, the lower the R-value. The heat that leaks through the wall is quickly transmitted to the moving air and carried away. The outer surface is cooled, drawing more heat through the wall. These heat losses can be reduced by wind screens or plantings that prevent fast-moving air from hitting the building skin.

Air infiltration heat losses through openings in buildings and through cracks around doors and windows are not easy to calculate because they vary greatly with tightness of building construction and the weatherstripping of windows, doors, and other openings. Small openings such as holes around outside electrical outlets or hose faucets can channel large amounts of cold air into heated rooms. Every intersection of one building material with another can be a potential crack if care isn't taken during construction. This is why, in home construction today, air/vapor barriers of 6-mil polyethylene sheets are commonly (and carefully) installed around the warm side of the building frame. They slow the passage of warm air (and moisture vapor) from inside to outside. Air barriers, sheets of polyethylene fibers that allow vapor, but not air, to pass through, are also installed around the outside of many buildings before the siding is installed. They keep cold air from passing through cracks between materials—cold air that forces warm air out the leeward side of the building. In both cases, special care is also taken around doors and windows, between floors, and around electrical and plumbing penetrations, to seal against the infiltration of cold air. This cold air has to be heated to room temperature. In the following calculations, we assume that the general wall construction is air-tight, and that only the infiltration through windows and doors needs to be considered.

The magnitude of air infiltration through cracks around doors and windows is somewhat predictable. It depends upon wind speeds and upon the linear footage of cracks around each window or door, usually the perimeter of the opening. If the seal between a window frame and the wall is not airtight, you must also consider the length of this crack. From the table "Air Infiltration Through Windows," you can approximate the volume of air leakage (Q) per foot of crack. With the temperature difference (ΔT) be-

AIR INFILTRATION THROUGH WINDOWS

Window Type	Remarks	Air leakage (Q)[1] at Wind velocity (mph)				
		5	10	15	20	25
Double-hung wood sash	Average fitted[2] non-weatherstripped	7	21	39	59	80
	Average fitted[2] weatherstripped	4	13	24	36	49
	Poorly fitted[3] non-weatherstripped	27	69	111	154	199
	Poorly fitted[3] weatherstripped	6	19	34	51	71
Double-hung metal sash	Non-weatherstripped	20	47	74	104	137
	Weatherstripped	6	19	32	46	60
Rolled-section steel sash	Industrial pivoted[2]	52	108	176	244	304
	Residential casement[4]	14	32	52	76	100

1. Air leakage, Q, is measured in cu ft of air per ft of crack per hr.
2. Crack = 1/16 inch.   3. Crack = 3/32 inch.   4. Crack = 1/32 inch.

SOURCE: ASHRAE, *Handbook of Fundamentals.*

tween inside and outside, you can determine the amount of heat required to warm this air to room temperature (ΔH_{inf}):

$$\Delta H_{inf} = (c)(Q)(L)(h)(\Delta T)$$

where $c = 0.018$ Btu/(ft^3°F) is the heat capacity of air, L is the total crack length in feet, and h is the time span in hours.

With 10 mph winds beating aginst an average double-hung, non-weatherstripped, wood-sash window, the air leakage is 21 cubic feet per hour for each foot of crack. Assuming the total crack length is 16 feet and the temperature is 65°F inside and 40°F outside, the total infiltration heat loss during an eight-hour time span is:

$$\begin{aligned}\Delta H_{inf} &= 0.018(21)(16)(8)(65 - 40) \\ &= 1210 \text{ Btu}\end{aligned}$$

If the same window is weatherstripped (Q = 13 instead of 21), then the infiltration heat loss is 749 Btu over the same time span. You can make a multitude of other comparisons using the Q-values given in the table.

Apply the above formula to the total crack length for each different type of crack leakage. The total crack length varies with room layout: for rooms with one exposure, use the entire measured crack length; for rooms with two or more exposures, use the length of crack in the wall having most of the cracks; but in no case use less than one-half of the total crack length.

You can also use this formula to calculate the heat loss through infiltration under the worst, or ''design'' conditions your house will undergo. For these conditions, use the outdoor design temperatures and average wind speed for your area. Fortunately, the design temperature does not usually accompany the maximum wind speed. Average winter wind velocities are given for a number of localities in the *Climatic Atlas of the United States*.

The total seasonal heat loss through air infiltration is calculated by replacing $h \times \Delta T$ with the total number of degree hours, or 24 times the number of degree days:

$$\text{Seasonal } \Delta H_{inf} = c(Q)(L)(24)(D)$$

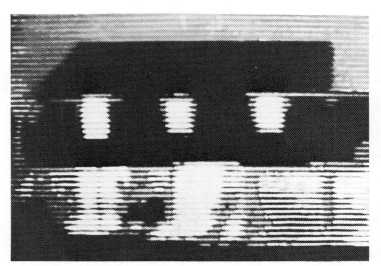

Infrared photographs showing thermal radiation from a conventional house. Note that more heat escapes from an uninsulated attic (top) than from an insulated one (bottom). SOURCE: Pacific Gas and Electric Co.

RADIATION HEAT FLOW

Radiation works together with conduction to accelerate heat flow through walls, windows, and roofs. If surrounding terrain and vegetation are colder than the outside surfaces of your house, there will be a net flow of thermal radiation to these surroundings. Your roof will also radiate substantial amounts of energy to the cold night sky. If the relative humidity is low, as much as 30 Btu per hour can be radiated to the sky per square foot of roof. This radiation can rapidly cool your roof surface to temperatures lower than the outside air temperature, thereby increasing the temperature difference across the roof section and the overall heat flow through the roof.

In summer, this radiative heat flow provides desirable nocturnal cooling, particularly in arid areas. In the winter, however, this nocturnal cooling is an undesirable effect. Well-insulated roofs are necessary to prevent excessive losses of heat.

If the interior surfaces of walls and windows are colder than the objects (and people!) inside a room, there will be a net flow of thermal radiation to these surfaces. A substantial flow of heat radiates to the inside surfaces of windows, which are much colder during winter than adjacent walls. This flow warms the inside surface of the glass, and more heat is pumped to the outside air because of the greater temperature difference across the glass. Extra glazing, special glazing, or window insulation can reduce this flow drastically.

In both examples above, radiation heat flow enhances the transfer of heat from warmer to cooler regions. Its effects are included in the calculation of conduction heat loss through surfaces of the house. But don't ignore radiation heat flow when taking preventive measures.

Heat Load Calculations

So far, you have learned to calculate the heat losses through the individual surfaces and cracks of a house. To calculate the overall heat loss (or heat load) of a house, you merely sum the losses through all surfaces and cracks. The heat load of a house depends on its construction and insulation and varies with the outside temperature and wind velocity.

To indicate just how bad things can get, let's use a drafty, uninsulated, wood-frame house as an example. Assume it's 40 feet long and 30 feet wide. It has uninsulated stud walls and a hardwood floor above a ventilated crawl space. The low-sloped ceiling has acoustical tile but is otherwise uninsulated, under a roof of plywood and asphalt shingles. The house has eight single-pane, double-hung, wood-sash windows (each 4 feet high by 2.5 feet wide) and two solid oak doors (each 7 feet by 3 feet).

First we need the U-values of each surface. From the "Sample Calculations of U-values" given earlier in this chapter, we know that an uninsulated stud wall has a U-value of 0.27. From the appendix, we get U = 1.13 for single-pane windows, and R = 0.91 for one inch of oak. Adding the resistance of the inside and outside air films, we get:

$$R_t = 0.68 + 0.91 + 0.17 = 1.76 \text{ or}$$
$$U = 1/1.76 = 0.57$$

for the doors.

The calculation of the U-values of the floor and ceiling is a bit more involved. The hardwood floor has three layers—interior hardwood finish (R = 0.68), felt (R = 0.06), and wood subfloor (R = 0.98)—and essentially still air films above and below (R = 0.61 each). The resistances of all five layers are added to give

$$R_t = 2.94, \text{ or } U = 1/2.94 = 0.34.$$

About half the floor area is covered by carpets (an additional R = 1.23 including the rubber pad), and this half has a U-value of 0.24. The total resistance of the ceiling and roof is the sum of the resistances of eight different layers, including the acoustical tile (R = 1.19), gypsum board (R = 0.45), rafter air space (R = 0.80), plywood (R = 0.62), building paper (R = 0.12), asphalt shingles (R = 0.44), and the inside and outside air films (R = 0.62 and 0.17). These add to R_t = 4.41, and the U-value of the ceiling is U = 1/4.41 = 0.23.

For a 1°F temperature difference between indoor and outdoor air, the conduction heat loss

HEAT LOAD CALCULATIONS

			Conduction heat losses	
Surface	Area (ft^2)	U-value Btu/(hr ft^2 $^\circ$F)	1°F temp diff Btu/(hr $^\circ$F)	35°F outside Btu/hr
Walls	998	0.27	269	8,084
Windows	80	1.13	90	2,712
Doors	42	0.57	24	718
Bare floor	600	0.34	204	6,120
Carpeted floor	600	0.24	144	4,320
Ceiling	1200	0.23	276	8,280
Total Conduction Heat Losses			1,007	30,234

			Infiltration Heat Losses	
Crack around:	Length (ft)	Q-value (ft^2 hr ft)	1°F temp diff Btu/(hr $^\circ$F)	35°F outside Btu/hr
Window sash	62	111	124	3,716
Door	20	220	79	2,376
Window & Door frames	82	11	16	487
Total Infiltration Heat Losses			219	6,579

All calculations assume 15 mph wind.

through each surface is the product of the area of the surface times the U-value of the surface. If the design temperature is 35°F, for example, we multiply by (65 − 35) to get the design heat loss through that surface. The conduction heat losses through all surfaces are summarized in the table.

Infiltration heat losses are calculated using Q-values from the table "Air Infiltration Through Windows." Poorly fitted double-hung wood-sash windows have a Q-value of 111 in a 15 mph wind. Assume that around poorly fitted doors, the infiltration rate is twice that: 220 ft^3/hr for each crack foot. Also assume that there is still some infiltration through cracks around window and door frames as well, with a Q-value of 11.

These Q-values are then multiplied by the heat capacity of a cubic foot of air [0.018 Btu/(ft^3 °F)] and the total length of each type of crack to get the infiltration heat loss. Only windows and doors on two sides of the house (that is, four windows and one door) are used to get total crack lengths. The infiltration heat losses through all cracks are also summarized in the table.

In a 15 mph wind, the conduction heat loss of this house is 1007 Btu/hr for a 1°F temperature difference between indoor and outdoor air. Under the same conditions, the infiltration loss is 219 Btu/hr, or a total heat load of of 1226 Btu/(hr °F). Over an entire day, the house loses 24 (hours) times 1226 (Btu per house) for each 1°F temperature difference, or 29,424 Btu per degree day. Under design conditions of 35°F and a 15 mph wind, the heat load of this house is 36,813 Btu/hr (30, 234 + 6,579). The furnace has to crank out almost 37,000 Btu/hr to keep this house comfy during such times.

SEASONAL AND DESIGN HEAT LOADS

The total heat escaping from a house is the sum of the conduction heat loss and the convection heat loss through air infiltration, because the effects of radiative heat flow have already been included in these two contributions. The total conduction heat loss is itself the sum of conduction losses through all the exterior surfaces, including walls, windows, floors, roofs, skylights, and doors. The total conduction heat loss is generally one to four times the total convection heat loss through air infiltration, which includes all convection heat losses through cracks in walls and around windows and doors.

The ratio of the two losses depends heavily on the quality of construction. For example, the total conduction heat loss from a typical poorly insulated 1250 square feet house may be 1000 Btu/(hr °F) temperature difference between the inside and outside air, while the convection heat loss is only 250 Btu/(hr °F). If the temperature drops to 45°F on a typical winter night, the house loses a total of

$$1250(65 - 45) = 25,000 \text{ Btu/hr}$$

assuming the indoor temperature is 65°F.

The design temperatures introduced earlier allow us to estimate the maximum expected heat loss from a house. The design temperature for a locality is the lowest outdoor temperature likely to occur during winter. Houses are often rated in their thermal performance by the number of Btu per house that the heating system must produce to keep the building warm during these conditions. The design temperature for Oakland, California, is 35°F, so that

$$1250(65 - 35) = 37,500 \text{ Btu/hr}$$

is the *design heat load* that the heating system must be able to produce in the above house. The same house would have design heat loads of 62,500 Btu/hr in Chattanooga, Tennessee, where the design temperature is 15°F, and 98,750 Btu/hr in Sioux Falls, South Dakota, where the design temperature is −14°F. The cost to heat the house in Sioux Falls might persuade the owner to add some insulation!

Degree day information allows us to calculate the amount of heat a house loses in a single heating season. The greater the number of degree days for a particular location, the greater the total heat lost from a house. Typical homes lose 15,000 to 40,000 Btu per degree day, but energy conservation measures can cut these by more than half. Our example house loses $(24)(1250) = 30,000$ Btu per degree day, for example. If there are 2870 degree days, as in Oakland, California, the total heat loss over an entire heating season is 86.1 million Btu [(30,000)(2870)] or about 1230 therms (1 therm = 100,000 Btu) of gas burned at 70 percent efficiency [86.1/(100,000)(0.7)]. In most other regions of the country, where seasonal heat loads are much greater and energy costs higher, energy codes are more stringent.

II
Passive Solar Systems

As the position of the heavens with regard to a given tract on the earth leads naturally to different characteristics, owing to the inclination of the circle of the zodiac and the course of the sun, it is obvious that designs for houses ought similarly to conform to the nature of the country and the diversities of climate.

Vitruvius,
Ten Books on Architecture

Energy conservation is the first step in good shelter design. Only the house that loses heat begrudgingly can use sunlight to make up most of the loss. Some people might think it rather dull to let sunlight in through the windows and keep it there, but others delight in its simplicity. In fact, conserving the sun's energy can often be more challenging than inventing elaborate systems to capture it.

Nature uses simple designs to compensate for changes in solar radiation and temperature. Many flowers open and close with the rising and setting sun. Many animals find shelters to shield themselves from intense summer heat, and bury themselves in the earth to stay warm during the winter.

Primitive peoples took a hint or two from nature in order to design shelters and clothing. But as we learned to protect ourselves from the elements, we lost much of this intuitive understanding and appreciation of natural phenomena. We rely more on technology than nature and the two are often in direct conflict.

27

The earth's heat storage capacity and atmospheric greenhouse effect help to moderate temperatures on the surface. These temperatures fluctuate somewhat, but the earth's large heat storage capacity prevents it from cooling off too much at night and heating up too much during the day. The atmosphere slows thermal radition from the earth's surface, reducing the cooling process. Because of these phenomena, afternoon temperatures are warmer than morning, and summer temperatures reach their peak in July and August.

A shelter design should reflect similar principles. Weather variations from one hour to the next or from cold night hours to warm daytime hours should not affect a shelter's internal climate. Ideally, not even the wide extremes of summer and winter would affect it. There are countless examples of indigenous architecture based on these criteria. Perhaps the most familiar of these is the heavy adobe-walled homes of the Pueblo Indians. The thick walls of hardened clay absorb the sun's heat during the day and prevent it from penetrating the interior of the home. At night, the stored heat continues its migration into the interior, warming it as the temperatures in the desert plummet. The coolness of the night air is then stored in the walls and keeps the home cool during the hot day. In many climates houses made of stone, concrete, or similar heavy materials perform in a like fashion.

A shelter should moderate extremes of temperature that occur both daily and seasonally. Caves, for example, have relatively constant temperatures and humidities year round. Likewise, you can protect a house from seasonal temperature variations by berming earth against the outside walls or molding the structure of the house to the side of a hill.

On sunny winter days, you should be able to open a house up to the sun's heat. At night, you should be able to close out the cold and keep this heat in. In the summer, you should be able to do just the opposite: during the day close it off to the sun, but at night open it up to release heat into the cool night air.

The best way to use the sun for heating is to have the house collect the sun's energy itself, without adding a solar collector. To achieve this, a house must be designed as a *total solar heating system* and meet three basic requirements:

The house must be a heat trap. It must be well insulated against heat loss and cold air infiltration. There's no point in making the house a solar collector if the house isn't energy-conserving. This is done with insulation, weatherstripping, shutters, and storm windows, or special glazings.

The house must be a solar collector. It must use direct-gain systems to let the sunlight in when it needs heat and keep it out when it doesn't; it must also let coolness in as needed. These feats may be accomplished by orienting and designing the house to let the sun penetrate the living space during the winter and by using shading to keep it out during the summer.

The house must be a heat storehouse. It must store the heat for times when the sun isn't shining. Houses built with heavy materials such as stone and concrete do this best.

28

3

The House as a Solar Heating System

The best way of using the sun's energy to heat a house is to let it penetrate directly through the roof, walls and windows. You should attempt to maximize your heat gain from insolation during cold periods, and minimize it during hot weather. You can do this with the color of your house, its orientation and shape, the placement of windows, and the use of shading.

Traditionally, solar heat gains have not entered into the computation of seasonal heating supply or demand. Unfortunately, most of the research done on solar gain applied to hot weather conditions and to reducing the energy required for cooling. But all that changed in the early 1980s. Still, the data that apply to heating are difficult to understand and difficult to use in building design. This chapter is an attempt to translate these data into useful design tools.

ORIENTATION AND SHAPE

Since solar radiation strikes surfaces oriented in different directions, with varying intensity, a house will benefit if its walls and roofs are oriented to receive this heat in the winter and block it in the summer. After much detailed study of this matter, a number of researchers have reached the same conclusion that primitive peoples have always known: the principal facade of a house should face within 30 degrees of due south (between south-southeast and south-southwest), with due south being preferred. With this orientation, the south-facing walls can absorb the most radiation from the low winter sun, while the roofs, which can reject excess heat most easily, catch the brunt of the intense summer sun.

In his book *Design With Climate,* however, Victor Olgyay cautions against generalizing to all building locations. He promotes the use of ''sol-air temperatures'' to determine the optimal orientation. These temperatures recognize that solar radiation and outdoor air temperatures act together to influence the overall heat gain through the surfaces of a building. Because the outdoor air temperatures are lower in the morning and peak in the mid-afternoon, he suggests that a house be oriented somewhat east of due south to take advantage of the early morning sun when heat is needed most. In the summer, the principal heat gain comes in the afternoon, from the west and southwest, so the house should face *away* from this direction to minimize the solar heat gain in that season. Depending upon the relative needs for heating and cooling, as well as upon other factors (such as winds), the optimum orientation will vary for different regions and building sites. The accompanying diagram gives the best orientations for four typical U.S. climate zones, as determined by Olgyay's sol-air approach.

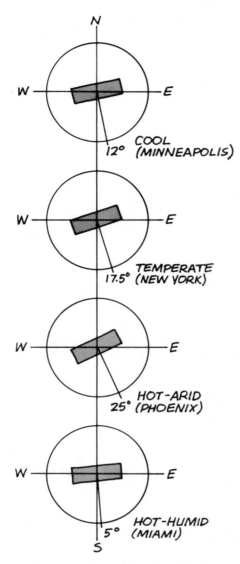

Optimum house orientations for four different U.S. climates.

East and west sides receive 2.5 times more in summer than they do in winter.
• At lower latitudes (less than 35°N) houses gain even more on their south sides in the winter than in the summer. East and west walls can gain two to three times more heat in summer than the south walls.
• The square house is not the optimum form in any location.
• All shapes elongated on the north-south axis work with less efficiency than the square house in both winter and summer. The optimum shape in every case is a form elongated along the east-west direction.

Of course, other factors influence the shape of a house, including local climate conditions (e.g., early morning fog), the demands of the site, and the needs of the inhabitants. But energy conservation can often be successfully integrated with these factors.

The relative insolation for houses with various shapes, sizes, and orientations can be a very useful aid at the design stage, particularly for placement of the windows. The first chart shown here lists the relative insolation for different combinations of house shape, orientation, and floor and wall area. Values in this chart are for January 21, and are based on the next chart, "Solar Heat Gain Factors for 40°N Latitude." The ASHRAE *Handbook of Fundamentals* provides similar information for many other latitudes. These factors represent the clear day solar heat gain through a single layer of clear, double-strength glass. But they can be used to estimate the insolation on the walls of a house.

From the relative solar insolation data, you may note that a house with its long axis oriented east-west has the greatest potential for total solar heat gain, significantly greater than that for a house oriented north-south. The poorest shape is the square oriented NNE-SSW or ENE-WSW. In doubling the ground floor area, the optimal east-west gain increases by about 40 percent because the perimeter increases by 40 percent. If you doubled the floor area of a house by

A house also benefits in solar heat gain because of different ratios of length to width to height. The ideal shape loses the minimum amount of heat and gains the maximum amount of insolation in the winter, and does just the reverse in the summer. Olgyay has noted that:

• In the upper latitudes (greater than 40°N), south sides of houses receive nearly twice as much solar radiation in winter as in summer.

FACADE ORIENTATIONS		INSOLATION ON WALL (Btu/day)				
		a	b	c	d	Total
N	A	118	508	1630	508	2764
	B	84	722	1160	722	2668
	C	168	361	2320	361	3210
	DOUBLE B	118	1016	1630	1016	3780
	DOUBLE C	236	508	3260	508	4512
N 22½°	A	123	828	1490	265	2706
	B	87	1180	1060	376	2703
	C	174	590	2120	188	3072
	DOUBLE B	123	1656	1490	530	3799
	DOUBLE C	246	828	2980	265	4319
N 45°	A	127	1174	1174	127	2602
	B	90	1670	835	180	2775
	C	180	835	1670	90	2775
	DOUBLE B	127	2348	1174	254	3903
	DOUBLE C	254	1174	2348	127	3903
N 67½°	A	265	1490	828	123	2706
	B	188	2120	590	174	3072
	C	376	1060	1180	87	2703
	DOUBLE B	265	2980	828	246	4319
	DOUBLE C	530	1490	1656	123	3799

BUILDING SIZES: RELATIVE WALL AND FLOOR AREAS.

Variation A — 1f, 1 w × 1 w

Variation B or C — 1f, .71 w × 1.42 w

Variation double B or double C — 2f, 1 w × 2 w

Relative insolation on houses of different shape and orientation on January 21 at 40°N latitude. Listed values represent the insolation on a hypothetical house with w = 1 foot. To get the daily insolation on a house of similar shape with w = 100 feet, multiply these numbers by 100.

adding a second floor, the wall area and the total solar insolation would double.

This study does not account for the color of the walls, the solar impact on the roof, the variations in window location and sizes, or the effects of heat loss. A detailed analysis would also include the actual weather conditions. However, this study does produce relative values to help you make preliminary choices.

COLOR

The color of the roofs and walls strongly affects the amount of heat which penetrates the house, since dark colors absorb much more sunlight than light colors do. Color is particularly im-

portant when little or no insulation is used, but it has less effect as the insulation is increased. Ideally, you should paint your house with a substance that turns black in winter and white in summer. In warm and hot climates, the exterior surfaces on which the sun shines during the summer should be light in color. In cool and cold climates, use dark surfaces facing the sun to increase the solar heat gain.

Two properties of surface materials, their *absorptance* (represented by the Greek letter alpha, α), and *emittance* (represented by the Greek letter epsilon, ϵ), can help you estimate their radiative heat transfer qualities. The *absorptance* of a surface is a measure of its tendency to absorb sunlight. *Emittance* gauges its ability

31

SOLAR HEAT GAIN FACTORS FOR 40° N LATITUDE, WHOLE DAY TOTALS
Btu(ft^2 day): Values for 21st of each month

	Jan	Feb	Mar	Apr	May	Jun	Jul	Aug	Sep	Oct	Nov	Dec
N	118	162	224	306	406	**484**	422	322	232	166	122	98
NNE	123	200	300	400	550	**700**	550	400	300	200	123	100
NE	127	225	422	654	813	**894**	821	656	416	226	132	103
ENE	265	439	691	911	1043	**1108**	1041	903	666	431	260	205
E	508	715	961	1115	1173	*1200*	1163	1090	920	694	504	430
ESE	828	1011	1182	*1218*	*1191*	1179	*1175*	*1188*	1131	971	815	748
SE	1174	1285	**1318**	1199	1068	1007	1047	1163	1266	1234	1151	1104
SSE	1490	**1509**	1376	1081	848	761	831	1049	1326	1454	1462	1430
S	*1630*	*1626*	*1384*	978	712	622	694	942	*1344*	*1566*	*1596*	*1482*
SSW	1490	**1509**	1370	1081	848	761	831	1049	1326	1454	1462	1430
SW	1174	1285	**1318**	1199	1068	1007	1047	1163	1266	1234	1151	1104
WSW	828	1011	1182	*1218*	1191	1179	*1175*	*1188*	1131	971	815	748
W	508	715	961	1115	1173	*1200*	1163	1090	920	694	504	430
WNW	265	439	691	911	1043	**1108**	1041	903	666	431	260	205
NW	127	225	422	658	813	**894**	821	656	416	226	132	103
NNW	123	200	300	400	550	**700**	550	400	300	200	123	100
HOR	706	1092	1528	1924	2166	**2242**	2148	1890	1476	1070	706	564

Figures in **bold** type: Month of highest gain for given orientations.
Figures in *italic:* Orientations of highest gain in given month.
Figures in ***bold italic***: Both month and orientation of highest gains.

SOURCE: ASHRAE, *Handbook of Fundamentals.*

Absorptance, Reflectance, and Emittance

Sunlight striking a surface is either absorbed or reflected. The absorptance (α) of the surface is the ratio of the solar energy absorbed to the solar energy striking that surface: $\alpha = I_a/I$, where I_a is absorbed solar energy and I is incident solar energy. A hypothetical "blackbody" has an absorptance of 1—it absorbs all the radiation hitting it, and would be totally black to our eyes.

But all real substances reflect some portion of the sunlight hitting them, even if only a few percent. The reflectance (ρ) of a surface is the ratio of solar energy reflected to that striking it: $\rho = I_r/I$, where I_r is reflected solar energy and I is incident solar energy. A hypothetical blackbody has a reflectance of 0. The sum of α and ρ for opaque surfaces is always 1.

All warm bodies emit thermal radiation, some better than others. The emittance (ϵ) of a material is the ratio of thermal energy being radiated by that material to the thermal energy radiated by a blackbody at that same temper-ature: $\epsilon = R/R_b$, where R is radiation from the material and R_b is radiation from the blackbody. Therefore, a blackbody has an emittance of 1.

The possible values of α, ρ and ϵ lie in a range from 0 to 1. Values for a few common surface materials are listed in the accompanying table. More extensive listings can be found in the appendix under "Absorptances and Emittances of Materials."

The values listed in this table (and those in the appendix) will help you compare the response of various materials and surfaces to solar and thermal radiation. For example, flat black paint (with $\alpha = 0.96$) will absorb 96 percent of the incoming sunlight. But green paint (with $\alpha = 0.50$) will absorb only 50 percent. Both parts (with emittances of 0.88 and 0.90) emit thermal radiation at about the same rate if they are at the same temperature. Thus, black paint (with a higher value of α/ϵ) is a better absorber of sunlight and will become hotter when exposed to the sun.

32

to emit thermal radiation. These properties are explained further in the sidebar given here, and sample values of α and ϵ are listed in the table. Also listed in the table are the materials' reflectance values (represented by the Greek letter rho, ρ).

Substances with large values of α are good absorbers of sunlight; those with large values of ϵ are good emitters of thermal radiation. Substances with a small value of α, particularly those with a small value of α/ϵ, like white paint, are good for surfaces that will be exposed to the hot summer sun (your roof and east and west walls, for example). Those that have a large value of α, particularly those with large α/ϵ, like black paint, are good for south-facing surfaces, which you want to absorb as much winter sunlight as possible.

ABSORPTANCE, REFLECTANCE, AND EMITTANCE OF MATERIALS

Material	Absorptance	Reflectance	Emittance	Absorptance/ Emittance
White plaster	0.07	0.93	0.91	0.08
Fresh snow	0.13	0.87	0.82	0.16
White paint	0.20	0.80	0.91	0.22
White enamel	0.35	0.65	0.90	0.39
Green paint	0.50	0.50	0.90	0.56
Red brick	0.55	0.45	0.92	0.60
Concrete	0.60	0.40	0.88	0.68
Grey paint	0.75	0.25	0.95	0.79
Red paint	0.74	0.26	0.90	0.82
Dry sand	0.82	0.18	0.90	0.91
Green roll roofing	0.88	0.12	0.94	0.94
Water	0.94	0.06	0.96	0.98
Black tar paper	0.93	0.07	0.93	1.00
Flat black paint	0.96	0.04	0.88	1.09
Granite	0.55	0.45	0.44	1.25
Graphite	0.78	0.22	0.41	1.90
Aluminum foil	0.15	0.85	0.05	3.00
Galvanized steel	0.65	0.35	0.13	5.00

4

Conservation First:
The House as a Heat Trap

If you design a house to collect and store solar heat, you should design the house to hold that heat. The escape of heat from a house during winter is usually called its "heat loss." In addition, houses also absorb heat through their walls and windows during summer—their "solar heat gain." Retarding this movement of heat both into and out of a house is the essence of energy conservation in housing design. Fortunately, most efforts to reduce winter heat loss also help reduce summer heat gain.

There are three primary ways that heat escapes from a house: (1) by conduction through walls, roofs, and floors, (2) by conduction through windows and doors, and (3) by convection of air through openings in the exterior surface. Conduction works together with radiation and convection—within the walls and floors and at the inner and outer wall surfaces—to produce the overall heat flow. The third mode of heat loss includes air infiltration through open windows, doors, or vents, through penetrations in the building "envelope," and through cracks in the skin of the house or around windows and doors.

Depending upon the insulation of the house, the number and placement of windows, and the movement of air, the ratios of the three modes to the total can vary widely. If the total heat loss is divided evenly among these modes, and

any one mode is reduced by half, the total heat loss is reduced by only one sixth. Clearly, you should attack *all three modes* of heat loss with the same vigor if you want to achieve the best results.

AIR INFILTRATION

People require some outdoor air for ventilation and a feeling of freshness, and the penetration of air through the cracks in the surface of a house usually satisfies this need—particularly if cigarette smoking is avoided. You should make every effort, however, to reduce such uncontrolled air infiltration. As you reduce other heat loss factors, the penetration of outdoor air becomes a greater part of the remaining heat loss.

Air infiltration can account for 20 to 55 percent of the total heat loss in existing homes. In those with some insulation, the heat loss from air infiltratation exceeds conduction losses through the walls, ceiling, and floor by up to 25 percent. Insulating older homes often requires a major overhaul, and tackling infiltration problems is the logical place to begin.

Measures for reducing infiltration include general "tightening up" of the structure and foundation, caulking and weatherstripping the doors and windows, redesigning fireplace air flow, creating foyer or vestibule entrances, installing a vapor barrier in or on the walls, and creating windbreaks for the entrances and the entire house.

One of the main reasons historically for installing building paper between the plywood sheathing and the exterior siding of houses was to reduce air infiltration through cracks in the walls. However, the fragile material did little to reduce infiltration after being penetrated by siding nails. Today's energy-conserving homes have a polyethylene-fiber air barrier wrapped around the sheathing—a continuous sheet that blocks the infiltration of cold air, but not the exfiltration of moisture vapor. Good trim details on a house exterior also reduce air penetration. Mortar joints in brick and concrete block facades should be tight and complete.

To tighten up your house, start with the obvious defects. Close up cracks and holes in the foundation, and replace missing or broken shingles and siding. Hardened, cracked caulking on the outside of the house should be removed and replaced with fresh caulking. Be sure to refit obviously ill-fitting doors and windows. And plug up interior air leaks around moulding, baseboards and holes in the ceiling or floor. More important than cracks in the wall surfaces, however, are those around the windows and doors. Different types of windows vary greatly in their relative air infiltration heat losses, but weatherstripping improves the performance of any window, particularly in high winds, because it checks infiltration where the edges of doors and windows meet (or don't quite meet!) their frames.

Weatherstripping is readily available at the local hardware store. Different types are required for different applications. For example, gasket compression-type weatherstripping (with a fabric face, peel-off back, and adhesive coating) is best suited for hinged door and casement and awning type windows. For sliding windows, the spring bronze or felt-hair weatherstripping is more appropriate. Forget about the cheap, spongy stuff. Its effectiveness deteriorates rapidly and so does its appearance.

Fixed windows save the most energy. You need operable windows for ventilation, but how many do you really need? In an existing house some windows can usually be rope-caulked shut for the winter. Double-hung windows are more of a problem. Your best bet is to rely on storm windows or insulating shutters to reduce infiltration and conduction losses. Be sure to caulk around the perimeter of storm window frames.

Before you try to stem infiltration between windows or doors and their frames, make sure no air is leaking around the *outside* edges of the frames. Caulking will remedy any such problems and prevent water seepage during driving rainstorms. The caulking compounds with superior adhesive qualities are generally called sealants. It is worthwhile to obtain high quality sealants because caulking is a lot of work, and inferior compounds can decompose after one winter!

If you plan to install new windows, you should be careful in selecting them. Operable windows should be chosen for their tight fit when closed—not only when first installed, but also after being used hundreds of times over a period of decades. Pivoted and sliding windows are the loosest, and casement and awning windows are among the closest fitting.

One obvious energy conservation step is to close a fireplace when not in use. If the fireplace is old and doesn't have a damper, install one. You can get more heat from a fireplace by using a C-shaped tubular grate to cradle the burning wood. Cold air is drawn into the tubes at the bottom, warmed and delivered back into the room from the top by thermosiphoning. Fireplace insert packages are available with built-in vents, and glass doors that block room air from rising up the chimney. Another way to extract more heat is to install vents in the flue that reclaim heat from hot air rising up the chimney. You should also provide a fresh intake air inlet for the fireplace or woodstove to draw air directly from the basement or outdoors. Otherwise, the fire will continue to draw warm air from the rooms, effectively cooling the same living space you want to heat by causing cold air to be pulled in from the outside.

AIR QUALITY

Better energy-conserving building techniques and infiltration control can mean less fresh air passing through the house. Houses with less than 0.5 air changes per hour can have excessive levels of carbon dioxide and moisture from occupants and cooking; formaldehyde from plywood, furniture, carpets, and tobacco smoke; radon from soils and groundwater; and combustion gases from kerosene heater, woodstoves, gas appliances, and furnaces. Controlling indoor air pollution at each source is the best remedy. Choose building materials and furniture without urea-formaldehyde glues or foams. Vent combustion appliances and provide separate outside combustion air. Install fans (on timers) in kitchens and bathrooms. Seal cracks and openings around penetrations to the basement. These will take care of the problem in the average home.

In superinsulated houses with air-infiltration rates less than 0.5 air changes per hour, an air-to-air heat exchanger may be necessary. Heat exchangers remove the heat from stale exhaust air and transfer it to fresh intake air. Air-to-air heat exchangers can be centrally located or wall-mounted like an air-conditioner. Central heat exchangers can provide fresh air to all the rooms.

In average construction, an exhaust-only fan may suffice. As it exhausts air, the negative pressure it creates inside the building draws clean air in through tiny cracks around windows and between floors, but does nothing to recover heat from the exhausted air. In houses where indoor air pollution is controlled at each source, occasional opening of a window may be all that's necessary.

WIND CONTROL

Wind is the arch-culprit in the moment-to-moment variation of the amount of air that penetrates a house. Olgyay reports in *Design With Climate* that a 20 mph wind doubles the heat loss of a house normally exposed to 5 mph

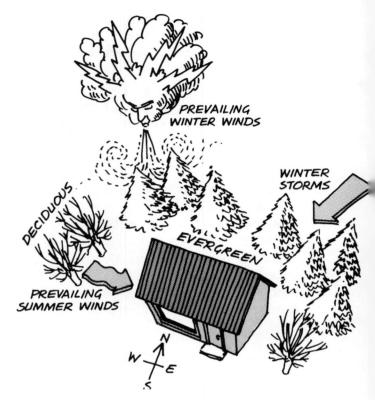

Proper orientation and vegetation shields protect a house from the wind.

winds. He also notes that the effectiveness of a belt of sheltering trees increases at higher wind velocities. With good wind protection on three sides, fuel savings can be as great as 30 percent.

Buildings should be oriented away from prevailing winter winds or screened by natural vegetation to block heat-pilfering air flows around windows and doors. Vegetation should be dense and eventually reach as high as the house. The distance from the house to the wind-break, measured from the home's leeward side, should not exceed five times the building height. Local agricultural extension services can suggest the trees and shrubs best suited to your climate and the appropriate planting distance from the house. Man-made windscreens, such as baffles, can also be very effective.

Winter winds blow, whistle, and wail from the north and west in most locales. Entrances should not be located on these sides, and the number of windows (the smaller the better!) should be kept to a minimum. Wind directions do vary, however, with locality and season. Monthly maps of the "Surface Wind Roses," in the *Climatic Atlas of the United States,* can be most helpful in the layout of windows and doors. These maps give the average wind velocity at many weather stations and show the percentage of each month that the wind blows in various directions. But remember that wind direction and speed depend very much on local terrain.

Plenty of cold outside air flows into a house every time you open a door, particularly if the door is on the windward side. However, a foyer or vestibule entrance can reduce this problem by creating an "air lock" effect. If a door opens into a hallway, another door can be positioned about four feet into that hallway to make a foyer. If your entrance has no hallway, you can build two walls out from either side of the door, add a roof, and install a second door. This addition can be simple and inexpensive—you needn't insulate the vestibule walls, just weatherstrip both doors. And be sure that the new door opens outwards for rapid exit in case of fire!

It may come as a surprise that air can infiltrate the walls themselves. Wind pressure forces air through the tiny cracks in the wall materials. A good vapor barrier will keep that cold air from reaching the living spaces while fulfilling its primary purpose, that of maintaining comfortable indoor humidity. For older homes with no such vapor barrier, installing one is only practical if the inside surface of the walls is being removed for extensive rehabilitation or remodeling. If your home needs a vapor barrier but you have no intention of ripping your walls apart, certain paints with low permeabilities are sold as vapor barriers.

AIR AND VAPOR BARRIERS

The daily activites of a family of four can produce two to three gallons of water vapor per day. Water vapor also flows into the building from basements and crawlspaces. Just as heat flows from areas of greater to lower temperatures (hot to cold), vapor migrates (or diffuses) from areas of greater to lower vapor pressures. It also travels in the air that infiltrates through cracks in walls, ceilings, and floors. Almost all of the moisture is carried by infiltration. Less than two percent moves by diffusion in a home with a typical vapor barrier and an infiltration rate of one air change per hour.

Moisture in the warm air will condense in the wall and ceiling cavities if it meets a cold surface. If enough of the moisture vapor condenses, it can saturate the insulation, reducing its R-value and eventually causing rot and decay. Vapor flow due to diffusion can be prevented with the use of a vapor barrier, such as polyethylene film or aluminum foil. Infiltration and exfiltration, which force moisture vapor through the building cracks and accounts for as much as 50 percent of the heat loss in a well-insulated home, can be slowed with a reasonably tight *air* barrier.

The following guidelines will help control condensation in homes:

• Avoid trapping moisture within a cavity. Use materials in the outer skin that are at least five times as permeable as the inner skin. Seal all cracks and joints. The air barrier should be as tight as possible.
• There should be at least twice as much insulation outside the vapor barrier as inside. In high-moisture areas, the vapor barrier should be on the warm side of all the insulation.
• Avoid any gaps in the insulation that could cause cold spots and result in condensation.

Condensation on double-glazing may indicate that you have problems in walls and ceilings that have inadequate vapor barriers. To prevent moisture infiltration from crawlspaces, vent them in all seasons except the dead of winter. Place a continuous 6–mil polyethylene moisture barrier across the floor of the crawlspace, and provide a tight vapor barrier above the floor insulation. If winter humidity levels

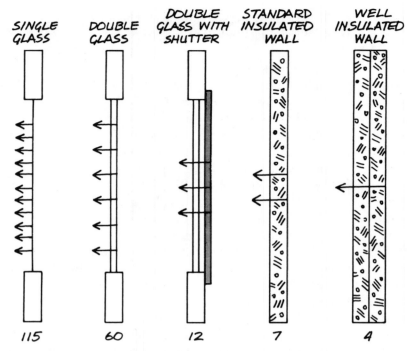

Relative heat losses through various types of windows and walls. These values represent only conduction heat loss.

inside the living space are kept below 40 percent, the potential for damage in wall and ceiling cavities will be limited.

Besides chimneys and flues, sources of infiltration that carry cold air in and moisture out are walls and basement (60 percent), windows and doors (20 percent), and ceilings (20 percent). Air enters through joints in materials in the building envelope and holes in the vapor barrier. To limit infiltration, use high-quality caulking to seal gaps between surfaces that do not move, such as where windows and framing, trim and siding, or sill and foundation meet. Weatherstrip all doors and windows. Air barriers—high density polyethylene fiber films stretched around the outside of the building frame—are also used to reduce infiltration. They allow moisture vapor to pass, but testing shows they can reduce infiltration from 35 to 47 percent in the average home.

As the infiltration rate is reduced below 0.5 air changes per hour, the relative humidity increases rapidly. The tighter the infiltration con-

trols, the more important the vapor barrier becomes. *Vapor* barriers—large thin sheets of transparent polyethylene around the inside of the building envelope—limit moisture migration. If you seal it very carefully at every seam and at window, door, plumbing, mechanical, and electrical penetrations, it will also serve as an *air* barrier. Installing two separate barriers, one *air* and one *vapor*, may be only slightly more expensive than installing one very tight vapor barrier, and they offer a greater defense and are easier to install. An alternative to the very tight polyethylene vapor barrier is caulking between the frame and the subfloor, and between the gypsum wallboard and the framing, and then painting the interior walls with a vapor barrier paint.

WINDOWS

The conduction heat losses through the surfaces of a house also increase with wind velocity. The lower the R-value of a surface, the more

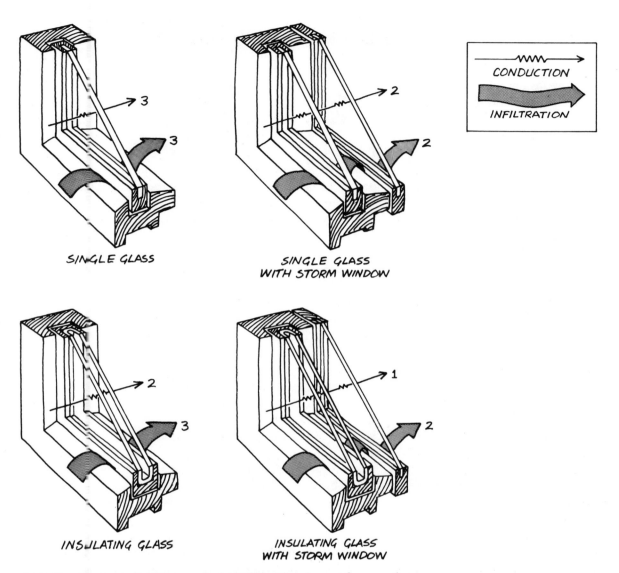

CONDUCTION

INFILTRATION

SINGLE GLASS

SINGLE GLASS
WITH STORM WINDOW

INSULATING GLASS

INSULATING GLASS
WITH STORM WINDOW

Relative heat conduction and air infiltration losses fom various windows. An added storm window cuts both kinds of heat loss.

you need to protect it from wind. A single-pane window or skylight needs much more wind protection than a well-insulated wall, because the air film clinging to its exterior surface contributes more to its overall thermal resistance. As the air film thickness decreases with the increase in the air velocity striking it, the effective insulating value of the film decreases. The decrease is large for doors and windows but almost negligible for well-insulated walls.

Various types of window and wall constructions differ widely in the amount of heat they transmit. Under the same indoor and outdoor air conditions, a single pane of glass will conduct 115 Btu, double glass will conduct 60 Btu, and a well-insulated wall will conduct only 4 Btu. You will lose the same quantity of heat through a well-insulated wall 30 feet long and 8 feet high as through a single glass window 2 feet wide and 4 feet high! A single-pane window loses heat about 20 times as quickly as a well-insulated wall of the same total area. There are a number of ways you can cut these losses, with high-performance glazings, storm windows, and window insulation.

In existing buildings, a storm window almost halves conduction heat loss for single-pane windows and also reduces air infiltration. A two-window sash (the standard single-pane window in combination with a storm window) can be superior to a single-sash insulating glass (because of the larger insulating air space) as long as it is sealed well around the perimeter. A standard window of double glazing in conjunction with a storm window is even better.

Insulating curtains are made of tightly woven material lined with loose stuffing, a blanket-type insulating material, or other heavy material. They are fitted at the top and bottom and travel in tracks at the sides to create a tight seal yet permit opening during times of winter sun. These curtains create a dead air space. With a reflecting layer on the outer surface, such insulating curtains can also be used to reduce solar heat gains in summer.

Another option is to insulate the windows with insulating shutters. Depending upon its thermal resistance, an insulating shutter can reduce conduction heat loss through a window by a factor ranging from two to ten. Shutters also reduce radiative heat transfer from warm bodies to the cold window glass and, depending upon construction, can practically eliminate air leakage. But window insulation can be very expensive, and it is not effective unless it is properly used more than 75 percent of the time. If its use cannot be guaranteed when operated manually, it should either have automatic controls that respond to light levels, or it should be passed up in favor of a special glazing, which works 24 hours a day without help.

HIGH-PERFORMANCE GLAZING

High-performance glazings are making their mark in new home construction, and may soon take the place of three layers of glass or two layers of glass with night insulation. These special windows are made to reduce heat loss. Their main advantage is that they cut heating bills over a 24-hour period.

There are only two approaches to improving the performance of a window: design it to transmit more light (i.e., heat) into the house, or manufacture it to lose less heat out. For years, low-iron glass was the solution to the first. Multiple glazings and window insulation solved the second.

In the 1970s, a vacuum-coated polyester film called Heat Mirror™ was introduced. Suspended between two panes of glass, the film allows less heat to escape by creating two insulating air spaces. But in addition, the coating itself is very selective in the wavelengths of radiation it transmits. Visible and near-infrared light pass through easily. But once changed to heat, the energy has a hard time passing back out because Heat Mirror™ reflects long-wave radiation (heat) back in. When placed between two layers of glass, it makes a lighter window than one made with three panes of glass (also cheaper to install), and has an R-value greater than 4.0.

Glass manufacturers took this vacuum-deposited metal oxide "sputtering" process a step further in the early 1980s and developed a process to "soft" coat glass. The coating is placed on an inner surface of an insulated glass unit. By being inside the sealed air space between the two layers of glass, the low-emissivity coating is protected from moisture, which can destroy it. These soft-coated windows have R-values greater than 2.0. (See table, "Glazing Properties.") That is slightly less than triple-glazed units, but the small difference is made up in lower costs and lighter weight.

In the mid-1980s, glass manufacturers developed a way to make a tougher coating that needs no special handling or protection. As the glass comes off the float line, the metal-oxide coating is sprayed onto the hot glass and becomes an integral "hard" coat as the glass cools. This pyrolitic process produces windows

GLAZING PROPERTIES

Glazing 1/8"	Air Space	Transmittance (Solar)	Shading Coefficient	Winter U-Value
Single	--	0.85	1.00	1.16
Double	1/4"	0.74	0.90	0.55
Triple	1/4"	0.61	0.85	0.39
Double, with low-e soft coat on outer surface of inner pane	1/4"	0.52	0.74	0.44
Double, with low-e hard coat on outer surface of inner pane	1/4"	0.51	0.83	0.52
Triple, with Heat Mirror between two glass panes	3/8"	0.46	0.62	0.25
Tripane, with anti-reflective film between two glass panes	3/8"	0.66	0.85	0.36
Quadpane, with two anti-reflective films between two glass panes	3/8"	0.63	0.82	0.26

with slightly lower R-values than the soft-coats, but a longer life makes them more attractive.

Another high-performance film increases window efficiency by transmitting more light. Anti-reflective glazings such as 3M Sungain™ have R-values of 3.85 when two layers are suspended in a double-glazed window called Quadpane™. These windows have a solar transmittance of 0.63—better than the quadrupled glass transmittance of 0.50. The units are also available with one layer of film called Tripane™. Added benefits of both low-emissivity and anti-reflective glazings are warmer interior glass surfaces for greater comfort and less condensation, lighter weight for easier installation, and less interior fabric fading because they block more ultraviolet light.

INSULATION

The only way further to reduce heat loss through air-tight walls, floors, and roofs is to add more resistance to this heat flow. Insulation retards the flow of heat, keeping the interior surfaces warmer in winter and cooler in summer. Because of radiation heat transfer from your body to the walls (which can be 8°F to 14°F colder than the room air during winter if the walls are poorly insulated), you can feel cold and uncomfortable even when the room air is 70°F. Eliminate this "cold-wall effect" by adding insulation and you will feel comfortable at lower thermostat settings.

Lowering the thermostat is the easiest way to reduce winter heating cost (but perhaps the most difficult for many of us to accept). The heat loss through walls and windows is proportional to the difference between indoor and outdoor temperatures. Reducing this difference can definitely reduce your heat loss. You can do this without undue discomfort by wearing an extra sweater, or by using more blankets while sleeping. The accompanying table shows that lowering the thermostat at night does save energy. A nightly 10°F setback reduces energy consumption by at least 10 percent in every city listed.

Well-insulated buildings also foster more uniform distribution of air temperatures. The air adjacent to cold, uninsulated walls cools, becomes more dense, and falls to the floor, displacing the warm air. These "ghost" drafts are considerably reduced in well-insulated houses. You can reduce the U-value of an exterior surface, and consequently its heat loss, by adding more insulation. However, your investment for insulation will quickly reach the point of diminishing return. For example, by adding two inches of polystyrene board insulation to the exterior of a concrete wall, you can reduce its U-value from 0.66 to 0.11, an 83 percent decrease in heat loss. Adding another two inches of polystyrene lowers the U-value to 0.06, an additional savings of only 7.5 percent. Your money may be better spent on extra window glazing and weatherstripping, depending on your climate.

R-VALUES OF COMMON INSULATORS

Insulation Material	R-Values		
	For one inch	Inside 2 x 4 stud wall*	Inside 2 x 6 stud wall*
Vermiculite	2.5	11.9	16.9
Mineral wool	3.0	13.7	19.7
Fiberglass	3.5	15.5	22.4
Polystyrene	4.0	17.2	25.2
Cellulose fiber	4.5	18.9	27.9
Urethane	6.5	25.9	38.9
Polyisocyanurate	7.0	27.7	41.7
Phenolic foam	8.3	32.2	48.8

* Includes insulating value of siding, sheathing, and air films, but not the effects of direct conduction through framing versus insulation.

SOURCE: E. Eccli, *Low-Cost Energy-Efficient Shelter*.

PERCENT FUEL SAVINGS WITH NIGHT THERMOSTAT SETBACK FROM 75° F
(8-hour setback: 10 pm to 6 am)

City	Setback		
	5°F	7.5°F	10°F
Atlanta	11	13	15
Boston	7	9	11
Buffalo	6	8	10
Chicago	7	9	11
Cincinnati	8	10	12
Cleveland	8	10	12
Dallas	11	13	15
Denver	7	9	11
Des Moines	7	9	11
Detroit	7	9	11
Kansas City	8	10	12
Los Angeles	12	14	16
Louisville	9	11	13
Milwaukee	6	8	10
Minneapolis	8	10	12
New York City	8	10	12
Omaha	7	9	11
Philadelphia	8	10	12
Pittsburg	7	9	11
Portland	9	11	13
Salt Lake City	7	9	11
San Francisco	10	12	14
St. Louis	8	10	12
Seattle	8	10	12
Washington, DC	9	11	13

SOURCE: Minneapolis-Honeywell Data.

The *placement* of insulation is also important. First, you should insulate roofs and the upper portions of walls. Warm air collects at the ceilings of rooms, producing a greater temperature difference there between indoor and outdoor air.

Six inches of fiberglass batt insulation (R-19) for roofs and 3.5 inches (R-11) for walls were once the standards in cold climates. These standards are being quickly accepted in mild climates and greatly upgraded to R-38 and R-19 in cold ones. In extreme cold, builders are installing R-60 roofs and R-25 to R-40 walls. These are the "superinsulated" buildings. Rigid board insulation installed outside the framing

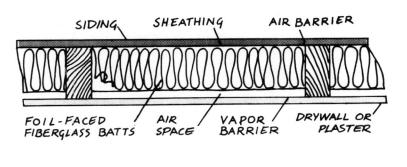

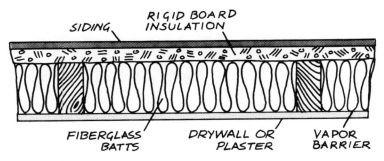

Adding rigid board insulation to exterior stud walls—plan view.

of conventional stud walls gives even better R-values because it slows the heat loss through the uninsulated wood frame, which can represent 15 to 25 percent of the total wall area. Construction details for adding such insulation are shown in the diagrams.

Attic insulation is the most crucial because substantial amounts of heat are lost in winter and gained in summer. An R-value of 20 to 30 can be obtained by applying thick blanket, batt, loose-fill, or poured insulation on the ceiling framing or directly on top of existing insulation. If the attic roof is too low, you can have a contractor install blown insulation. If neither is possible, wait until re-roofing time and add rigid board insulation.

The insulation of an existing stud wall is limited by the wall thickness. The only insulating materials that approach or exceed the desired resistance of 19 in a standard 2 × 4 stud wall are cellulose fiber or urethane. Mineral wool and fiberglass insulation won't do it (see table). Trying to compress these insulators to increase their resistance will have the opposite effect after a point—compression reduces the air spaces needed to slow the flow of heat. If you are fortunate enough to have 2 × 6 studs, you have many choices.

Cellulose fiber or polystyrene beads can be blown into wood frame walls, although holes will have to be bored in the interior wall between studs and above and below the firebreak. Later somone will have to patch the holes, providing you with an opportunity to use a vapor barrier paint. The other alternative—installing blanket or batt insulation—requires ripping out the interior walls. This makes sense only if the walls need replacing.

For masonry walls, one method is to blow loose-fill or foam insulation into the existing air spaces. This approach is possible if the plate for the ceiling rafters doesn't cover the concrete block cores or the cavity wall construction. Rigid board insulation can also be placed on the outside of a masonry wall and replastered or covered with siding.

Insulate floors and foundations last. Tacking foil-backed insulation supported by wire mesh to the underside of the floor (leaving a half inch air space) can provide a high resistance. If there isn't enough room to get under the house, seal the foundation but leave a few ventilation openings. For basements being used as living space, insulate the foundation walls all the way to the floor (interior) or footing (exterior).

5

Direct Gain Systems

Once the house is insulated well to retain solar or auxilary heat, it can be designed to act as a solar collector. Although the color, orientation, and shape of the house are important, the most significant factors in capturing the sun's energy are size and placement of windows. Openings in shelters are the origin of present-day windows: they were used for the passage of people and possessions, and for natural ventilation and lighting. These openings also allowed people to escape from indoor drudgeries by gazing off into sylvan surroundings. But the openings also had their discomforts and inconveniences. Animals and insects had free access, the inside temperature was difficult to regulate, and humidity and air cleanliness could not be controlled.

Although glass has been dated as early as 2300 BC, its use in windows did not occur until about the time of Christ. And only in the present century have the production and use of glass panes larger than eight or twelve inches on a side become possible. As the technology and economics improve, glass is replacing the traditional masonry or wood exterior wall. But the design problems accompanying this substitution have often been ignored or underrated.

Besides reducing the amount of electricity needed for lighting, glass exposed to sunlight captures heat through the greenhouse effect ex-

plained earlier. Glass readily transmits the short-wave visible radiation, but does not transmit the long-wave thermal radiation emitted after the light changes to heat when it hits an interior surface. Almost all this thermal radiation is absorbed in the glass and a substantial part of it is returned through radiation and convection to the interior space.

Experimental houses were built in the 1930s and 1940s with the major parts of south-facing walls made entirely of glass. The most extensive work with these "solar houses" was done by F.W. Hutchinson at Purdue University. In 1945, under a grant from Libbey-Owens-Ford Glass Company, he built two nearly identical houses. They were thermally and structurally the same, except that one house had a larger south-facing window area. Based on the performance of these two houses, Hutchinson reported that "the available solar gain for double windows in south walls in most cities in the U.S. is more than sufficient to offset the excess transmission loss through the glass."

Hutchinson also concluded that more than twice as much solar energy is transmitted through south-facing windows in winter than in summer. If the windows are shaded in summer, the difference is even greater. For a fixed latitude, the solar intensity does not vary strongly with the outside air temperature, but heat loss does.

SOLAR BENEFIT VALUES

City	Percent Possible Sunshine	Average Heating Season Temperature	Net Heat Gain Btu/(hr ft²) Single glass	Net Heat Gain Btu/(hr ft²) Double glass	City	Percent Possible Sunshine	Average Heating Season Temperature	Net Heat Gain Btu/(hr ft²) Single glass	Net Heat Gain Btu/(hr ft²) Double glass
Albany, NY	46	35.2	-12.8	5.6	Jacksonville, FL	40	62.0	13.9	18.1
Albuquerque, NM	77	47.0	18.0	30.2	Joliet, IL	53	40.8	2.9	12.8
Atlanta, GA	52	51.5	9.0	18.8	Lincoln, NB	61	37.0	-2.2	15.3
Baltimore, MD	55	43.8	2.0	15.9	Little Rock, AR	51	51.6	8.5	18.3
Birmingham, AL	51	53.8	10.9	19.5	Louisville, KY	51	45.3	1.5	14.6
Bismarck, ND	55	24.6	-20.1	4.0	Madison, WI	50	37.8	-7.6	9.5
Boise, ID	54	45.2	22.9	16.0	Minneapolis, MN	53	29.4	-15.7	5.8
Boston, MA	54	38.1	5.2	11.7	Newark, NJ	55	43.4	1.4	15.5
Burlington, VT	42	31.5	-19.5	.9	New Orleans, LA	37	61.6	11.7	16.1
Chattanooga, TN	50	49.8	5.9	16.7	Phoenix, AZ	59	59.5	21.9	27.5
Cheyenne, WY	67	41.3	5.7	20.9	Portland, ME	52	33.8	-7.2	12.0
Cleveland, OH	41	37.2	-13.7	3.7	Providence, RI	54	37.2	-6.1	11.3
Columbia, SC	51	54.0	11.2	19.6	Raleigh, NC	57	50.0	-10.0	20.6
Concord, NH	52	33.3	-12.0	7.4	Reno, NV	64	45.4	8.6	21.7
Dallas, TX	47	52.5	7.1	16.4	Richmond, VA	59	47.0	8.0	20.2
Davenport, IA	54	40.0	-3.1	12.8	St. Louis, MO	57	43.6	2.6	16.6
Denver, CO	70	38.9	5.2	21.7	Salt Lake City, UT	59	40.0	0.0	15.9
Detroit, MI	43	35.8	14.1	44.0	San Francisco, CA	62	54.2	17.3	25.7
Eugene, OR	44	50.2	2.7	13.2	Seattle, WA	34	46.3	-7.3	5.2
Harrisburg, PA	50	43.6	-1.5	12.5	Topeka, KS	61	42.3	3.8	18.4
Hartford, CT	53	42.8	-.3	14.1	Tulsa, OK	56	48.2	7.4	19.0
Helena, MT	52	40.7	-3.3	12.2	Vicksburg, MS	45	56.8	-10.7	17.7
Huron, SD	58	28.2	-14.1	8.0	Wheeling, WV	41	46.1	3.7	9.0
Indianapolis, IN	51	40.3	-4.6	11.2	Wilmington, DE	56	45.0	3.7	16.9

Consequently, the use of glass has greater potential for reducing winter heating demand in mild climates than in cold climates.

The table of "Solar Benefit Values" gives us plenty of evidence for this potential. Many of the cities studied showed net energy gains through single glass (a negative number represents a net loss), and all 48 cities studied showed net gains through double glass. The losses through single glass in some cities should be compared to the heat loss through a typical wall that the glass replaces.

There are a number of reasons that the quantity of solar energy that gets through a south window on a sunny day in winter is more than that received through that same window on a sunny day in summer.

1. There are more hours when the sun shines directly on a south window in winter than in summer. At 40°N latitude, for example, there are 14 hours of possible sunshine on July 21, but the sun remains north of east until 8:00 a.m. and goes to north of west at 4:00 p.m., so that direct sunshine occurs for only eight hours on

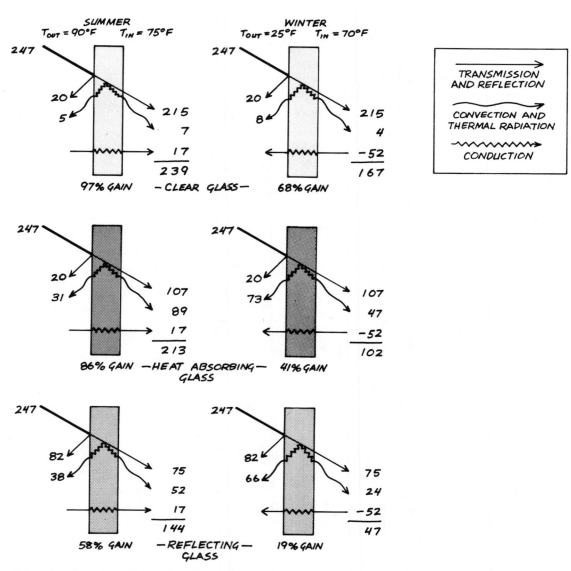

Solar heat gains through clear, heat-absorbing, and reflecting single glass. Listed values are in Btu per hour.

the south wall. But on January 21, the sun is shining on the south wall for the full ten hours that it is above the horizon.

2. The intensity of sunlight hitting a surface perpendicular to the sun's rays is about the same in summer and winter. The extra distance that the rays must travel through the atmosphere in the winter is offset by the sun's closer proximity to the earth in that season.

3. Since the sun is closer to the southern horizon during the winter, the rays strike the windows closer to perpendicular than they do in the summer when the sun is higher in the sky. This means less is reflected and more is transmitted. At 40°N latitude, 200 Btu strike a square foot of vertical window surface during an average hour on a sunny winter day, whereas 100 Btu is typical for an average summer hour.

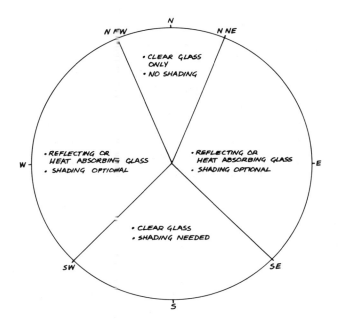

Different glass types are recommended for limiting summer heat gain for various window orientations.

In addition to these effects, the diffuse radiation from the winter sky is double that from the summer sky.

GLAZING

The type of glazing you use can have a significant effect on energy gains and losses. Single sheets of clear, heat-absorbing, and reflecting glass all lose about the same amount of heat by conduction. But there is a great difference in the amount of solar heat transmitted through different types of glass, as shown in the first table. The percentage summer and winter heat gains for single-glazed units of clear, heat-absorbing, and reflecting glass are summarized in the second table. The accompanying diagrams will give you an idea of the net heat gains for various combinations of single and double glass. The percentage of solar heat gain includes a contribution from heat conduction through the glass. The heat gains are approximate for the sunny day conditions shown, and no attempt has been made to account for the differing solar angles in summer and winter.

To reduce summer heat gain, you might use reflecting glass on the outside and clear glass on the inside of two-pane windows facing into the sun. Unfortunately, this combination drastically reduces the winter heat gain, and is not recommended for south-facing glass. Two clear panes of glass, low-emissivity double-glazed units (with the special coating on the outer surface of the inner pane), or anti-reflective triple- or quadruple-glazed units, are generally recommended for windows used for solar heat gain in winter. In either case, you must still use shading, natural and artificial, to keep out the hot summer sun.

In many climates, keeping the sunshine out during warm weather is very important to human comfort. In such areas, the use of special glazings is one alternative, especially for the east and west sides. The important factors to consider in the use of specialized glass bear repeating:

1. Such glass *does* reduce solar heat gain, which can be more of a disadvantage in the winter than an advantage in the summer.

2. Except for their higher insulating values, special glazings are almost always unnecessary on north, north-northeast, and north-northwest orientations. Reflecting and heat-absorbing glass only helps to control glare.

3. In latitudes south of 40°N, heat absorbing and reflecting glass should not be considered for south-facing windows.

4. The use of vegetation or movable shading devices is a more sensible solution than the use of heat-absorbing or reflecting glass for south, southeast, and southwest orientations.

SOLAR TRANSMITTANCE

Glazing Type	
Single, clear	0.85
Double, clear	0.74
Triple, clear	0.61
Triple, low-e film	0.46
Quad, clear	0.50
Double, low-e coating	0.52
Triple, anti-reflective film	0.66
Quad, anti-reflective film	0.63

The New Solar Home Book

PERCENTAGE HEAT GAIN THROUGH CLEAR, HEAT-ABSORBING AND REFLECTIVE GLASS

Glass Type	Summer	Winter
Single Glazing		
Clear	97	68
Heat-absorbing[1]	86	41
Reflective[2]	58	19
Double Glazing		
Clear outside & inside	83	68
Clear outside/		
heat-absorbing inside	74	52
Clear outside/		
reflective inside	50	42
Heat-absorbing outside/		
clear inside	42	28
Reflective outside/		
heat-absorbing inside	31	17

1. Shading coefficient = 0.5.
2. Shading coefficient = 0.35.

The four (or more) sides of a building need not, and in fact should not, be identical in appearance. Substantial savings in heating and cooling costs will result from the use of well-insulated walls on the north, east and west. The few windows needed on these sides of the house for lighting and outdoor views should use the glazing methods advocated here. In most areas of the United States, double-glazed clear glass windows or high-performance glazings on the south sides provide the optimum winter heat gain.

SHADING

Through the intelligent use of shading, you can minimize the summer heat gain through your windows. Perhaps the simplest and most effective methods of shading use devices that are exterior to the house, such as overhangs or awnings. One difficulty with fixed overhangs is that the amount of shading follows the seasons of the sun rather than the climatic seasons. The middle of the summer for the sun is June 21, but the hottest times occur from the end of July to the middle of August. A fixed overhang designed for optimal shading on August 10 causes the same shadow on May 1. The overhang designed for optimal shanding on September 21, when the weather is still somewhat warm and solar heat gain is unwelcome, causes the same shading situation on March 21, when the weather is cooler and solar heat gain is most welcome.

Shading a south window with a fixed overhang (at solar noon).

Sizing Overhangs

Overhangs can be effective shades for large south-facing vertical window areas. How much shade you want and when you want it depends on the home's heating and cooling load. You can size an overhang by choosing what months you want shade and how much of the window you want shaded (e.g., all or half the window). The depth of the overhang (O) and how high it is separated from the window (S) are found with simple trigonometry:

$$O = H/(\tan A - \tan B)$$

$$S = D \tan B$$

where H is the height of the shadow (measured down from the bottom of the overhang), A is the summer noon profile angle, and B is the winter noon profile angle.

The profile angle is difficult to envision. The figure shows that it is the angle between the horizon and the sun's rays, in a vertical plane perpendicular to the window. The noon profile angle is equal to (90 − L + D), where L is the latitude of the site and D is the declination of the sun.

Let's say you lived at 40°N latitude, and you wanted full shade on a four-foot high window on June 21st and no shade on September 21st.

The table lists the declination angles for the 21st day of each month. In this case:

$$A = 90 - L + D = 90 - 40 + 23 = 73$$

$$B = 90 - 40 + 0 = 50$$

$$O = H/(\tan A - \tan B) \\ = 4/(\tan 73 - \tan 50) = 1.92$$

$$S = D \tan B = 1.92 \tan 50 = 2.29$$

In this case, the overhang would need to be almost two feet deep and its lower edge would be over two feet above the window. If you could accept full shade on June 21st, but no shade on December 21st (and hence some shading on September 21st), the overhang could be shallower and closer to the top of the window:

$$A = 90 - 40 + 23 = 73$$

$$B = 90 - 40 - 23 = 27$$

$$O = 4/(\tan 73 - \tan 27) \\ = 1.45 \text{ ft (deep)}$$

$$S = 1.45 \tan 27 \\ = 0.74 \text{ ft (8 in) above the window.}$$

MONTHLY SOLAR DECLINATIONS

Month (Day 21)	Declination
December	-23
January/November	-20
February/October	-10
March/September	0
April/August	+11.6
May/July	+20
June	+23

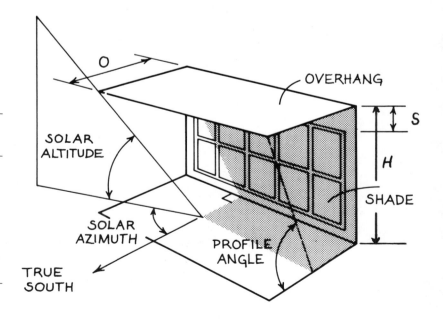

Vegetation, which follows the climatic seasons quite closely, can provide better shading year round. On March 21, for example, there are no leaves on most plants, and sunlight will pass readily (except through oak trees, which do not lose their leaves until late fall). On September 21, however, the leaves are still full, providing the necessary shading. Placement of deciduous trees directly in front of south-facing windows can provide shade from the intense midday summer sun. But watch out for trees with dense, thick branches that still shade even without their leaves. Even better is an overhanging trellis with a climbing vine that sheds its leaves in winter. Unfortunately, stalks remain and produce considerable shading in the winter as well, so the vines must be cut back in the fall.

Movable shading devices are even more amenable to human comfort needs than fixed overhangs or vegetation, but they have their own problems. Movable shading devices on the outsides of buildings are difficult to maintain and can deteriorate rapidly. Awnings are perhaps the simplest and most reliable movable shading devices, but their aesthetic appeal is limited. The requirement for frequent human intervention is often seen as a drawback. Operable shading placed between two layers of glass is not as effective as an exterior device, but it is still more effective than an interior shading device.

UV-TRANSMITTANCE

Glazing Type

Single, clear	0.78
Double, clear	0.64
Triple, clear	0.51
Triple, low-e film	0.43
Quad, clear	0.41
Double, low-e coating	0.29
Triple, anti-reflective film	0.06
Quad, anti-reflective film	0.02

Mini-blinds between glass panes can be expensive. Interior shading devices, such as roller shades and draperies, give the least effective shading but offer versatile operation by the people inside. And they do keep direct sunlight from bleaching the colors of walls, furniture, and floors. (The high-performance glazings are also effective in reducing this bleaching effect because they block more ultraviolet light than ordinary glass. The table lists the ultraviolet transmittance of different glazings.)

East- and west-facing glass is extremely difficult to shade because the sun is low in the sky both early morning and late afternoon. Overhangs do not prevent the penetration of the sun during the summer much more than they do during the winter. Vertical louvers or extensions are probably the best means of shading such glass, but you might consider reflecting and heat-absorbing glass or high-performance glazings. For this purpose, you should be familiar with the values of the *shading coefficient* of the various glasses. A single layer of clear, double-strength glass has a shading coefficient of 1. The shading coefficient for any other glazing system, in combination with shading devices, is the ratio of the solar heat gain through that system to the solar heat gain through the double-strength glass. Solar heat gain through a glazing system is the product of its shading coefficient times the solar heat gain factors. The solar heat factors for 40°N latitude were listed earlier. The ASHRAE *Handbook of Fundamentals* has a complete list for other latitudes.

SUN PATH DIAGRAMS

It is usually necessary to describe the position of the sun in order to determine the size of a window shading device. Earlier, we described the sun's path in terms of the solar altitude angle (θ) and the azimuth angle (ϕ). These can be determined for the 21st day of any month by

COEFFICIENTS FOR VARIOUS SHADING CONDITIONS

Condition	Coefficient
Clear double-strength glass, 1/8", unshaded	1.00
Clear plate glass, 1/4", unshaded	0.95
Clear insulating glass, two panes 1/8" plate, unshaded	0.90
Clear insulating glass, three panes 1/8" plate, unshaded	0.85
Double-glazed unit, 1/8" glass with low-e hard coat on surface 3, unshaded	0.84
Clear insulating glass, two panes 1/4" plate, unshaded	0.83
Clear insulating glass, three panes 1/4" plate, unshaded	0.78
Double-glazed unit, 1/8" glass with low-e soft coat on surface 3, unshaded	0.72
Clear glass with dark interior draperies	0.69
Heat-absorbing 1/4" plate glass, unshaded	0.68
Triple-glazed unit, 1/8" glass with low-e soft coat on surface 5, unshaded	0.67
Triple-glazed unit, 1/8" glass with low-e suspended film, unshaded	0.62
Blue reflective 1/4" glass, unshaded	0.58
Clear glass with light interior venetian blinds	0.55
Heavy-duty grey heat-absorbing 1/2" glass, unshaded	0.50
Heavy-duty grey heat-absorbing 1/2" glass with dark interior drapes (or medium venetian blinds)	0.42
Silver reflective 1/4" glass, unshaded	0.23
Silver reflective 1/4" glass with interior drapes or venetian blinds	0.19
Clear glass with exterior shading device	0.12

using tables, or can be calculated directly from formulas. Another method for determining solar altitude and azimuth for the 21st day of each month is the use of *sun path diagrams*. A different diagram is required for each latitude, although interpolation between graphs is reasonably accurate. Diagrams for latitudes from 24°N to 52°N are provided in the appendix. The 40°N diagram is reprinted here as an example.

You can also use sun path diagrams to determine the effects of shading devices. There are two basic categories of shading—a horizontal overhang above the window or vertical fins to the sides. As shown in the diagram, the shading angles *a* and *b* of these two basic obstructions are the two important variables available to the designer. The broader the overhang or fin, the larger the corresponding angle.

Each basic shading device determines a specific *shading mask*. A horizontal overhang determines a "segmental" shading mask while

vertical fins determine a "radial" one. These shading masks are constructed with the help of the shading mask protractor provided here. These masks can then be superimposed upon the appropriate sun path diagram for your latitude to determine the amount of shading on a window. Those parts of the diagram that are covered by the shading mask indicate the months of the year (and the times of day) when the window will be in shade.

Sun path diagrams and the shading mask protractor can also be used to design shading devices. If you specify the times of year that shading is needed and plot these on the appropriate sun path diagram, you have determined the shading mask for your desired condition. The shading angles *a* and *b* can be read from this mask using the shading mask protractor. From these angles you can then figure the dimensions of the appropriate shading devices.

Use of Sun Path Diagrams

A sun path diagram is a projection of the sky vault, just as a world map is a projection of the globe. The paths of the sun across the sky are recorded as lines superimposed on a grid that represents the solar angles. Sun path diagrams can be used to determine these angles for any date and time. Different sun path diagrams are needed for different latitudes.

As an example, find the solar altitude and azimuth angles at 4:00 p.m. on April 21 in Philadelphia (40°N). First locate the April line—the dark line running left to right and numbered "IV" for the fourth month—and the 4:00 p.m. line—the dark line running vertically and numbered "4." The intersection of these lines indicates the solar position at that time and day. Solar altitude is read from the concentric circles—in this case it's 30 degrees. The

solar azimuth is read from the radial lines—in this case it's 80 degrees west of true south. If you trust your judgement, you can also use these diagrams to give you the solar positions on days other than the 21st of each month.

The shading mask protractor provided here will help you to construct masks for any shading situation. First determine the shading angle of the horizontal overhang or vertical fins, as shown in the figure. For a horizontal overhang, find the arc corresponding to angle "a" in the lower half of the shading mask protractor. All the area above that arc is the segmental shading mask for that overhang. For vertical fins, find the radial lines corresponding to the shading angle "b" in the upper half of the shading mask protractor. All the area below these lines is the radial shading mask for those fins.

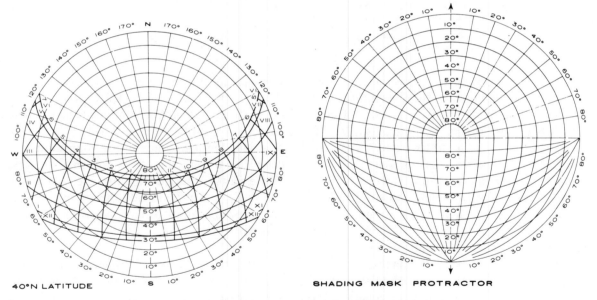

40°N LATITUDE   SHADING MASK PROTRACTOR

SOURCE: Ramsey and Sleeper, *Architectural Graphic Standards*.

HORIZONTAL OVERHANG

VERTICAL FINS

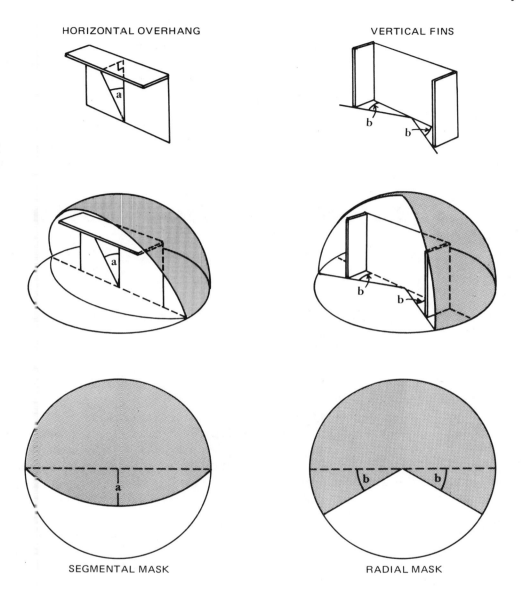

SEGMENTAL MASK

RADIAL MASK

Determining masks for horizontal and vertical shading obstructions. Use the shading mask protractor to convert a particular shading angle into the corresponding mask.

6

The House as a Heat Storehouse

A vital question in a solar-heated house is where to store the heat. When the house is used as the solar collector, it needs a method of "soaking up" or storing heat so it doesn't become too hot when the sun is shining, and retains some of this heat to use when it isn't. Probably the most efficient heat storage container is the material of the house itself—the walls, floors, and roofs. All materials absorb and store heat as they are warmed. For example, water or stone will absorb more heat for a fixed temperature rise than straw or wood. Heavy materials can store large quantities of heat without becoming too hot. When temperatures around them drop, the stored heat is released and the materials themselves cool down.

This *heat storage capacity* of various materials can be used to store the sun's heat for later use. Solar energy penetrates through walls, roofs, and windows to the interior of a house. This solar heat is absorbed in the air and surrounding materials. The air in the house is likely to heat up first. It then distributes this heat to the rest of the materials via convection. If they have already reached the temperature of the air or cannot absorb the heat quickly, the air continues to warm and overheats. The greater the heat storage capacity of the materials in the house, the longer it will take for the air to reach uncomfortable temperatures and the more heat can be stored inside the house.

If it is cold outside when the sun sets, the house begins losing heat through its exterior skin, even if it is well insulated. To maintain comfortable temperatures, this heat must be replaced. In houses which have not stored much solar heat during the day, auxiliary heating devices must provide this heat.

If the interiors are massive enough, however, and the solar energy has been allowed to penetrate and warm them during the day, the house can be heated by the sun, even at night. As the inside air cools, the warmed materials replace this lost heat, keeping the rooms warm and cozy. Depending upon the heat storage capacities of the inside materials, the amount of solar energy penetrating into the house, and the heat loss of the house, temperatures can remain comfortable for many hours. Really massive houses can stay warm for a few days without needing auxiliary heat from fires or furnaces.

During the summer, a massive house can also store coolness during the night for use during the hot day. At night, when outside air is cooler than it is during the day, ventilation of that air into the house will cool the air and all of the materials inside. Since they will be cool at the beginning of the next day, they can absorb and store more heat before they themselves become warm—cooling the indoor air as they absorb heat from it. Thus, if the materials are cool in the morning, it will be a long time before they have warmed to the point that additional cooling is needed to remove the excess heat.

TEMPERATURE SWINGS

The effects of varying outdoor temperatures upon the indoor temperatures can be very different for different types of houses. The first graph shows the effects of a sharp drop in outdoor temperature on the indoor temperatures of three types. Of the three, a lightweight wood-frame house cools off the fastest. It has little heat stored in its materials to replace the heat lost to the outside. A massive structure built of concrete, brick, or stone maintains its temperature over a longer period of time if it is insulated on the outside of the walls. The heavy materials which store most of the heat are poor insulators, and they must be located within the confines of the insulation.

A massive house set into the side of a hill or covered with earth has an even slower response to a drop in the outdoor air temperature. Ideally, the interior concrete or stone walls in this house are insulated from the earth by rigid board insulation. One or two walls can be exposed to the outside air and still the temperature will drop very slowly to a temperature close to that of the earth.

The second graph shows the effects of a sharp rise in outdoor temperature on the same three houses. Again, the lightweight house responds the fastest to the change in outdoor temperature; in spite of being well-insulated, its temperature rises quickly. The heavy houses, however, absorb the heat and delay the indoor temperature rise. The house set into a hill or covered with earth has the longest time delay in its response to the outdoor air change; if properly designed, it may never become too warm.

The effects of alternately rising and falling outdoor air temperatures on indoor air temperatures are illustrated in the third graph. Without any sources of internal heat the inside air temperature of the lightweight house fluctuates widely, while that of the earth-embedded house remains almost constant near the temperature

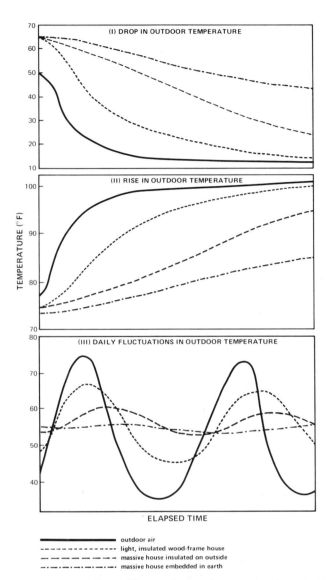

Effects of changes in outdoor air temperature on the indoor air temperatures of various houses.

of the earth. We say that massive houses, whose indoor temperatures do not respond quickly to fluctuations of outdoor temperature, have a large *thermal mass,* or *thermal inertia.*

If a house responds slowly to outdoor temperature fluctuation, you don't need heavy duty auxiliary equipment to keep the place comfortable. Although the furnace in a lightweight, uninsulated wood-frame house might not be used much on a cold, sunny day, it might have to labor at full throttle to keep the house warm at night. The massive earth-embedded house, on the other hand, averages the outdoor temperature fluctuations over a span of several days or even weeks. A bantamweight heating system (such as a wood stove, for example) could operate constantly to assure an even comfort level throughout the house.

HEAT STORAGE CAPACITIES

All materials vary in their ability to store heat. One measure of this ability is the *specific heat* of a material, which is the number of Btu required to raise one pound of the material 1°F. For example, water has a specific heat of 1.0, which means that 1 Btu is required to raise the temperature of 1 pound of water 1°F. Since one gallon of water weighs 8.4 pounds, it requires 8.4 Btu to raise it 1°F.

Different materials absorb different amounts of heat while undergoing the same temperature rise. While it takes 100 Btu to heat 100 pounds of water 1°F, it takes only 22.5 Btu to heat 100 pounds of aluminum 1°F. (The specific heat of aluminum is 0.225.) The specific heats of various building materials and other common materials found inside buildings are listed in the accompanying table.

The *heat capacity,* or the amount of heat needed to raise one cubic foot of the material 1°F, is also listed along with the density of each material. Although the specific heat of concrete, for example, is only one-fourth that of water, its heat capacity is more than half that of water. The density of concrete compensates somewhat for its low specific heat, and concrete stores relatively large amounts of heat per unit volume. As heat storage devices, concrete or stone walls insulated on the outside are superior to wood-framed walls having a plywood exterior and a gypsum wallboard interior with fiberglass insulation stuffed between them.

BUILDING WITH THERMAL MASS

Thermal mass is one of the most underrated aspects of current building practice. Unfortunately, heavy buildings are hardly the favorite children of architects and building contractors, because the visual weight of buildings is an important aesthetic consideration. Well-insulated homes with reasonable amounts of south glazing (no more than six percent of the floor area) usually have enough thermal mass in the standard building materials without adding more. Extra thermal mass is now looked at more as an "option" than a "necessity."

Massive fireplaces, interior partitions of brick or adobe, and even several inches of concrete or brick on the floor can greatly increase the

SPECIFIC HEATS AND HEAT CAPACITIES OF COMMON MATERIALS

Material	Specific Heat Btu/(lb °F)	Density lb/ft^3	Heat Capacity Btu/(ft^3 °F)
Water (40°F)	1.00	62.5	62.5
Steel	0.12	489	58.7
Cast iron	0.12	450	54.0
Copper	0.092	556	51.2
Aluminum	0.214	171	36.6
Basalt	0.20	180	36.0
Marble	0.21	162	34.0
Concrete	0.22	144	31.7
Asphalt	0.22	132	29.0
Ice (32°F)	0.487	57.5	28.0
Glass	0.18	154	27.7
White oak	0.57	47	26.8
Brick	0.20	123	24.6
Limestone	0.217	103	22.4
Gypsum	0.26	78	20.3
Sand	0.191	94.6	18.1
White pine	0.67	27	18.1
White fir	0.65	27	17.6
Clay	0.22	63	13.9
Asbestos wool	0.20	36	7.2
Glass wool	0.157	3.25	0.51
Air (75°F)	0.24	0.075	0.018

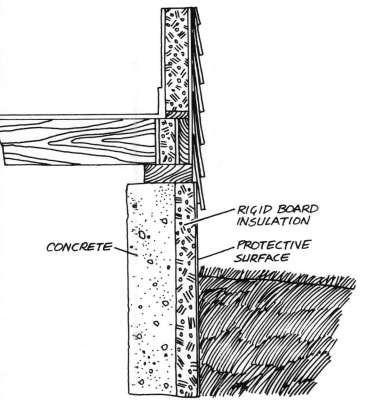

thermal mass of a house. Placing containers of water within the building confines, especially in front of a window, is a simple solution.

Putting insulation on the outside of a house is not standard construction practice and involves some new problems. Insulation has customarily been placed between the inner and outer surfaces of a wall. Insulation on the outside of a concrete or masonry wall requires protection from the weather and contact with people or animals.

In the example shown, three inches of rigid board insulation covers the outside surface of a poured concrete foundation. Above the surface of the ground, this insulation must be protected from rain, physical abuse, and solar radiation —particularly ultraviolet rays. Below ground level, it must be protected from the unmerciful attacks of moisture and vermin. The insulation could be placed inside the formwork before the concrete is poured, and the bond between the two materials would be extremely strong. But the insulation must still be protected above ground

Insulation on the exterior of a house must be protected from weather and vermin to at least one foot below grade.

Storing Heat in a Concrete Slab

Consider a 20 × 40 foot house with a well-insulated concrete slab 9 inches thick. By 5 p.m. on January 21, the slab has warmed up to 75°F from sunlight flooding in the south windows. From that time until early the next morning, the outdoor temperature averages 25°F, while the indoor air averages 65°F. If the house is well-insulated and loses heat at a rate of 300 Btu/(hr °F) and there is no source of auxiliary heat, what is the temperature of the slab at 9 o'clock the next morning?

The total heat lost from the house during that period is the product of the rate of heat loss (UA), times the number of hours (h), times the average temperature difference between the indoor and outdoor air (ΔT), or

$$\Delta H = (UA)\ (h)(\Delta T)$$
$$= 300(16)(65 - 25)$$
$$= 192,000\ Btu.$$

With a total volume of 600 cubic feet (20)(40)(0.75) and a heat capacity of 32 Btu/ (ft^3 °F), the concrete slab stores 19,200 Btu for a 1°F rise in its temperature. For a 1°F drop in its temperature, the slab releases the same 19,200 Btu. If the slab drops 10°F, from 75°F to 65°F, it will release just enough heat to replace that lost by the house during the night. So the slab drops to a temperature of 65°F by 9:00 the next morning.

In reality, things are a bit more complicated. But this exercise helps to give a rough idea of how much heat you can store in a concrete slab. If the house has 200 square feet of south windows, and a solar heat gain of 1000 Btu/ft^2 is typical for a sunny January day, the slab can store the 200,000 Btu of solar energy with a temperature rise of about 10°F. The stored solar heat is then released at night to keep the house warm as the inhabitants sleep.

level. The most popular alternative is to plaster the insulation with a "cementitious" material such as fiberglass-reinforced mortar.

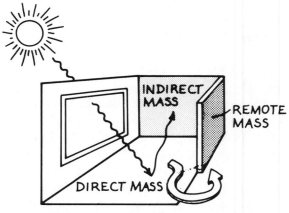

The three kinds of thermal mass, based on location.

SIZING MASS

Thermal mass in a building stores heat and releases it later to the space when the air temperature around it begins to drop. When sized properly, the mass can prevent overheating on a sunny afternoon, and can keep the auxiliary heat from turning on until later in the evening.

There are three ways mass can be "charged," that is, heated up. If it is "direct" mass, it is directly hit by the sun for at least six hours a day. If it is "indirect" mass, the sunlight hits another surface first and is reflected onto the mass. If it is "remote" mass, it is charged by warm air that flows by its surface through natural or forced convection. Direct, indirect, and remote mass all charge and discharge on one side only, as shown in the figure.

The effectiveness of the mass is directly related to its location. Direct mass stores more heat than the same surface area of indirect mass, and much more than remote mass. When you design, try to place the mass where the sun

strikes so that you can get by with as little mass as possible. This saves on costs.

The table shows the surface area of the three types of mass needed for each square foot of south glazing, for different materials of various thicknesses. The mass must be directly exposed to the room. Mass in the floor doesn't count if it's covered with carpet. Concrete or brick walls don't count as mass if they are hidden behind a frame wall.

In a well-insulated, light-frame house (R-19 walls, R-38 ceiling, triple-glazing), the building materials themselves are enough thermal mass to allow six to seven square feet of south glazing for every 100 square feet of heated floor area. In the average house (R-11 walls, R-19 ceiling, double-glazed), the building material's thermal mass allows 11 to 14 square feet of south glazing for every 100 square feet of heated floor area. To have more glazing than that, thermal mass must be added to the building to avoid overheating.

MASS SIZING: SQUARE FEET OF MASS NEEDED FOR EACH SQUARE FOOT OF SOUTH GLAZING

	Material Thickness	Direct Mass	Indirect Mass	Remote Mass
Concrete	4"	4	7	14
	6"	3	5	14
	8"	3	5	15
Brick	2"	8	15	20
	4"	5	9	18
	8"	5	10	19
Gypsum board	0.5"	76	114	114
	1"	38	57	57
Hardwood	1"	17	28	32
Softwood	1"	21	36	39

7
Indirect Gain Systems

A grasp of the principles of thermosiphoning—where the natural bouyancy of heated air or water is used to circulate heat—is crucial to an understanding of the passive uses of solar energy. When heated, air expands and becomes lighter than the surrounding air. The heated air drifts upward and cooler air moves in to replace it.

You have observed the process of thermosiphoning, also called natural convection, at work in a fireplace. Because the hot air just above the fire is much lighter than the surrounding air, it rises rapidly up the chimney. Cooler, heavier room air replaces it, bringing more oxygen to maintain the flames. Most of the fire's heat is delivered to the outdoors by this "chimney effect." Thermosiphoning is also a strong force in passive solar heating systems.

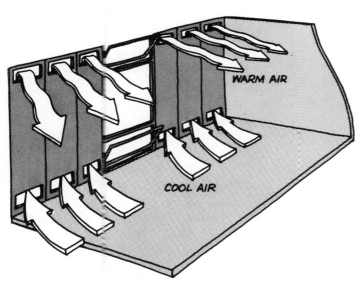

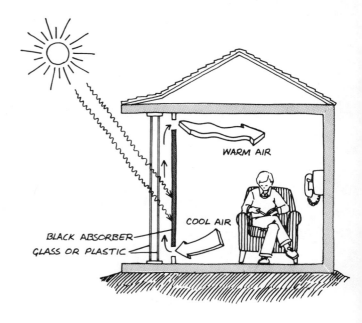

A thermosiphoning air collector—two views.

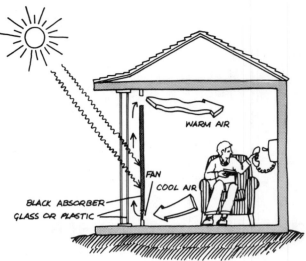

Added fan provides heat control.

Undesirable reverse thermosiphoning cools the room.

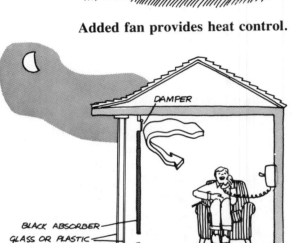

Damper prevents reverse thermosiphoning.

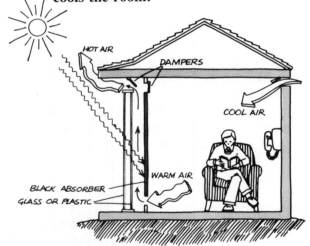

Chimney effect induces natural ventilation in summer.

THERMOSIPHONING AIR PANELS

The simplest form of a thermosiphoning air panel, or TAP, is illustrated in the diagrams. The air in the space between the glass and the blackened absorber wall is heated. It expands and becomes lighter, rises through the collector, and flows into the room from a vent at the top. Cool room air is drawn through another vent at the base of the wall, heated in turn, and returned to the room at the top. This process continues as long as there is enough sunlight to push the temperature of the absorber wall above the room air temperature.

To provide greater control of air flows, you can add a fan to the supply duct of the solar collector. Faster movement of air across the absorber surface boosts the collector efficiency and allows the use of a smaller air gap between absorber and glass. These hybrid collectors are called forced air panels, or FAPs. They differ from active air collectors only in their smaller scale and their dependence on the thermal mass of the building itself to prevent the space from overheating. A fan can also deliver warm air to other parts of the house, such as north rooms, or heat storage bins (in which case, they are active systems). Using fans with a proper combination of windows and wall collectors, you can simultaneously heat the rooms exposed to the sun and those in the shade.

Dampers help to control the air flow and prevent the cooling effect of *reverse* thermosiphoning. When the sun isn't shining, the air in the collector loses heat by conduction through the glass and radiation to the outside. As this air cools, it travels down the absorber face and flows out into the room. Warm room air is drawn in at the top to the collector and cooled in turn. Although this reverse thermosiphoning could be a benefit in summer, it is most undesirable in winter. It can be prevented by shutting dampers at night.

Dampers can operate manually or automatically. Natural air currents or fan pressure can open or close them. You can also use dampers in summer to prevent overheating by inducing natural ventilation through houses. Cool air can be drawn into the house from the north side and warm air expelled by the "chimney" exhaust system shown here. As with all air-type solar heating devices, dampers should be simple in design and operation. They should close tightly and there should be as few of them as possible.

TAP VARIATIONS

A number of variations on the basic design of TAPs can improve their performance. These variations include insulation, improved absorber surfaces, and dampers and fans to regulate the flow of air.

During a sunny winter day, no insulation is needed between the back of the absorber and

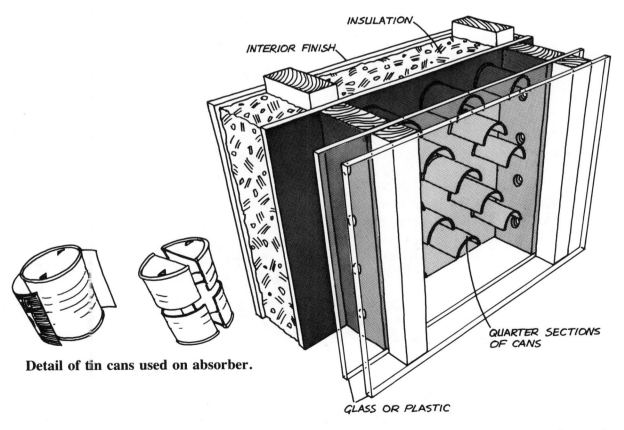

Detail of tin cans used on absorber.

Low-cost thermosiphoning air collector built onto an exterior wall.

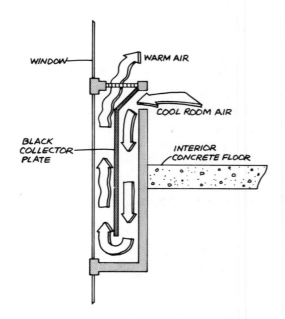

Cross-section of CNRS wall collector.

the room. To reduce room heat losses on cloudy days or at night, however, the wall should be adequately insulated behind the absorber.

A metal absorber plate isn't an absolute necessity for a TAP. Since the temperature of the collector wall does not get extremely high, blackened masonry or wood surfaces are also possible, and costs need not be excessive. Alternatives that increase the total absorber surface can be particularly effective, if they do not hinder the natural convection air flow. Rough surfaces make better absorbers than smooth ones. Pebbles cast in a blackened concrete wall are a good example of such an absorber surface. Special absorber sheets, made of a selective surface with an adhesive back, can greatly improve TAP performance.

Another option (shown in the figure) has tin cans cut into quarters and mounted on the standard plywood sheathing of conventionally-framed houses.

Some of the most significant work in thermosiphoning air collectors was done at the *Centre National de la Recherche Scientifique* (CNRS) in Odeillo, France. Under the direction of Professor Felix Trombe, this laboratory developed

several low technology approaches to solar heating. The main building remains an excellent example of the passive use of solar energy. Its south, east, and west walls are a composite of windows and thermosiphoning air panels, which supply about half of the building's winter heat. The TAPs are installed below the windows, between floors so that the view to the exterior isn't blocked.

A cross-sectional view of these collectors is shown in the accompanying diagram. Blackened corrugated metal sheets are located behind a single pane of glass. Solar radiation passes through the glass and is absorbed by the metal, which is contained entirely within the volume defined by the glass and duct. As the metal heats, so does the air between the absorber plate and the glass. The heated air flows upward through vents into the rooms. Simultaneously, cooler room air falls through a lower vent and sinks down between the back of the absorber and the duct wall. This air returns to the face of the absorber where it, too, is heated and expelled into the rooms.

No provision has been made to store the solar heat, other than the thermal mass of the building itself—particularly the reinforced concrete slab floors. Consequently, the system is most effective when the sun is shining—almost 90 percent of the daytime hours in Odeillo. The air temperature in the offices and laboratories remains relatively constant during the day. Even during February, auxiliary heat is required only at night and on overcast days. Outdoor temperatures are relatively cool in summer, allowing the use of east and west facing collectors, which would overheat most buildings in hot climates.

MASS WALLS

Heat storage capacity can be added directly to TAP vertical wall collectors. The overall simplicity of this synthesis of collector and heat storage is compelling. Large cost reductions are possible by avoiding heat transport systems of ducts, pipes, fans, and pumps. Operation and

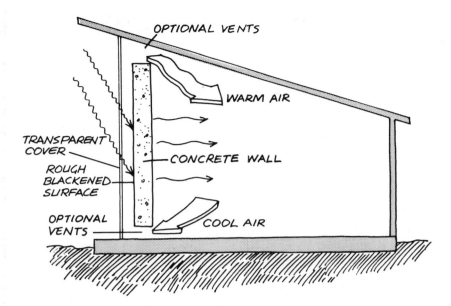

OPTIONAL VENTS

WARM AIR

TRANSPARENT COVER

CONCRETE WALL

ROUGH BLACKENED SURFACE

OPTIONAL VENTS

COOL AIR

A concrete mass wall collector. Solar collection, heat storage, and heat distribution are combined in one unit.

maintenance are far simpler, and comfort and efficiency generally greater, than collectors with remote storage.

The schematic diagram shows a concrete wall used as a solar collector and heat storage device. When sunlight strikes the rough blackened surface, the concrete becomes warm and heats the air in the space between wall and glass. Some of the solar heat is carried off by the air, which rises and enters the room, but a large portion of this heat migrates slowly through the concrete. The wall continues to radiate heat into the house well into the night, after the thermosiphoning action has ceased. In energy-conserving buildings with proper insulation levels and infiltration control, mass walls can be sized to maintain comfort for two or three days of sunless weather.

Mass walls are usually constructed without the vents for spaces used primarily at night. All day long, the sun's heat is driven through the thick wall, until it reaches the inside surface at the end of the day. The thicker the wall, the longer it will take for the heat can be conducted through the wall. Once it reaches the inside surface, the heat is delivered to the living space through radiation and convection currents.

MASS WALL VARIATIONS

A number of variations in the design of concrete walls is possible. The wall can be constructed from poured concrete, or hollow masonry blocks filled with sand or concrete. Empty voids can be used as air ducts for thermosiphoning. Brick or adobe can also be used instead of concrete, and need not be painted black if dark enough.

There are advantages in making the space between the concrete wall and the glass covers wide enough for human use. The space can be used as a porch or vestibule, or even as a greenhouse. But the thermosiphoning heat flow to the interior does not work very well for such large spaces because the air does not get quite as hot. Fortunately, there will still be large heat flows by conduction through the wall and radiation to the rooms.

Mass walls can also be constructed of other materials, using water or phase-change materials to store the heat. Special containers for water walls and whole manufactured units for phase-change materials are available.

The pioneering work in mass walls was done at Odeillo under the direction of Professor Trombe and architect Jacques Michel. Thus mass walls

63

made of concrete are often referred to as Trombe (pronounced Trohm) walls. The first buildings were two four-room houses, each with a floor area of 818 square feet and a collector area of 516 square feet. The collectors operate in a fashion similar to the one in the previous diagram, except that to prevent reverse thermosiphoning, the lower ducts are located above the bottom of the collector. Cool air settling to the bottom at night is trapped there.

Because they are not very well insulated, the houses lose about 22,000 Btu per degree day. Nevertheless, the concrete wall collectors supply 60 to 70 percent of the heat needed during an average Odeillo winter, where temperatures frequently plummet to 0°F. From November to February, the collectors harvest more than 30 percent of the sunlight falling upon them. Over a typical heating season, this passive system supplies about 200,000 Btu (or the usable heat equivalent of 2 gallons of oil) per square foot of collector.

WALL, WINDOW, AND ROOF COLLECTORS

Ease of construction is perhaps the most important reason for the emphasis on vertical wall collectors rather than sloping roof collectors. Glazing is much easier to install, weatherproof, and maintain in a vertical orientation. The cost difference between windows and skylights is testimony to this fact. Builders estimate you can install three windows for every skylight at about the same cost. It is much easier to keep weather out of vertical surfaces than tilted or horizontal ones. There are fewer structural complications with walls than with roofs, and you needn't worry much about hail or snow build-up. Another important architectural constraint of large, steeply-pitched roofs is that interior space under such roofs is difficult to use.

The total amount of clear day solar heat gain on south walls follows the seasonal need very closely. In most of the United States, the greatest heat gain on vertical south walls occurs in

December and January, the coldest months, and the least occurs in June and July. The midwinter clear day insolation on vertical south walls is only about 10 percent less than that on tilted roofs facing south. With an additional 10-50 percent more sunlight reflected onto vertical surfaces from fallen snow, they can actually receive more solar heat gain than tilted surfaces. Other types of reflectors, such as swimming pools, white gravel, and concrete walks, work well with vertical collectors. South walls can be shaded easily in summer, preventing the collector surface from reaching high temperatures.

At first glance, it seems foolish to remove a window which admits light and heat directly, only to replace it with an opaque wall solar collector. But don't forget the advantages of a mix of windows and collectors: direct gain, indirect gain, view, ventilation, and egress. Interior wall surface is lost if the entire south facade is glass, and excessive sunlight can dam-

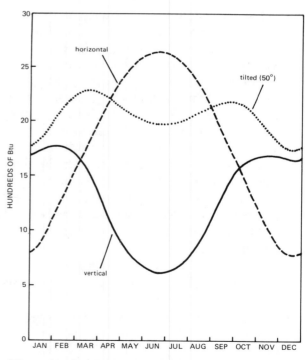

Clear day insolation on horizontal surfaces, and on south-facing vertical and tilted surfaces. Reflected radiation not included.

age furniture, floors, and fabrics. A section of wall provides an interior space where you may place delicate objects that could not take direct sunlight. People also can be very uncomfortable when the sun shines directly on them. Overheating is often a problem with an all-glass wall, even with massive floors and partitions. But with solar collectors and heat storage in the south walls, the excess heat can be transported to cooler parts of the house or trapped and stored for later use.

SUNSPACES

Sunspaces are the modern solar equivalent of attached greenhouses. Originated in Roman times to meet the demands of Tiberius Caesar for fresh cucumbers out of season, greenhouses were used in 19th-century Europe as both a supply of year-round food and a source of winter heat from the sun. Called "conservatories" in England, they were built in sizes ranging from small window units to room-sized structures.

Today, site-built or prefabricated sunspaces have become a very popular mode of capturing the benefits of passive solar heat and making winter living more cheerful. As a heat source, a sunspace with 100 square feet of glazing can, in one winter, offset up to 250 gallons of heating oil, 6560 kilowatts of electricity, or 3000 cubic feet of natural gas. As a greenhouse rather than primarily a heat source, a sunspace can work wonders on houseplants, and supply fresh vegetables all winter long. Or a sunspace can be used mainly as additional year-round living space. (In this case it is a direct-gain system and not truly a sunspace.) How the space is to be used will affect choices of glazing, heat storage materials, and heat transfer methods.

Glazings and Orientation

Sunspaces collect solar heat through south-facing glass or plastic glazing. Glass can last a long time—up to 50 years if no one throws a rock through it in the meantime. Using low-iron glass, with solar transmissivity greater than ordinary glass, will improve sunspace performance. Polycarbonate sheets, fiberglass-reinforced polyesters, and polyethylene glazings are other choices in sunspace glazing. They can be less expensive, lighter in weight, and less likely to break than glass, but they age more quickly, losing their strength and appearance.

If the sunspace will only be used as a solar collector, and closed off from the house at night, single glazing will do. Single glazing may also be suitable if you use insulated shutters or shades that store out of the way in the daytime and cover the glazing at night. Otherwise, double glazing shoud be used. (Triple-glazed glass is only worth the extra cost in severe climates.) New high-performance glazings are available for site-built sunspaces with a layer of high transmittance film sandwiched between two glass layers, or a low-emissivity coating on the inner pane of double glazing. They offer a high degree of protection from 24-hour heat loss.

A sunspace should face due south, but up to 15 degrees either east or west will have little effect on performance. Facing it slightly east offers desirable sunlight and warmth in the morning. The space should not be shaded during the day. Direct sun should be allowed in at least from 10:00 a.m. to 2:00 p.m., and more is better.

The best glazing angle for collecting winter sun is between 50 and 60 degrees. But sloped glazing is difficult to seal against leaks year-round, and is harder than vertical glazing to fit with window insulation in winter and shades in summer. Sunspaces with sloped glazing are colder at night in the winter and hotter all summer. Vertical glazing solves these problems, with only a 10 to 30 percent loss in efficiency. Ground reflectance from snow, gravel, or sidewalks can cut this loss in half.

Storing and Moving the Heat

If a sunspace is to be more than a solar collector, it will need thermal mass to retain the heat when the sun goes down. A brick or concrete floor works well. So do water-filled containers painted a dark color. Phase-change salts have four to five times the heat storage capacity of water, but are expensive and have often proved unreliable. Whatever is used should be positioned to capture direct and reflected solar radiation, and have as great a surface area as possible.

At least three square feet of concrete or brick floor (4 to 6 inches thick) or three gallons of water are required for every square foot of south glazing. Heat can also be stored in a rockbed under the sunspace and/or living room floor by blowing warm air from the sunspace through it.

There are several choices of how to connect the sunspace and the house. The two spaces can be open to one another so that the sun's heat penetrates directly. If you choose this design, use high-performance glazings or very good insulating shades over double-glazing at night, and insulate well around the rest of the structure, or heat loss will be high. (The sunspace would really be a direct-gain space in this case.)

Double-glazed sliding glass doors are a second option. Direct solar radiation can still penetrate the living space, and the view through the sunspace to the outside can be maintained. If the doors are left open during the day, warm air can freely pass to the living area. When the doors are closed at night, the sunspace acts as a thermal buffer, reducing heat losses in adjacent rooms.

A standard wood-frame wall with R-10 to R-15 insulation prevents solar radiation from passing through, but warm air can pass through open windows and doors. A solid masonry wall offers a combination of storage and distribution. If it is meant to warm only the sunspace, it should be six to eight inches thick and insulated on the house side. If the wall's stored heat is to be shared by the sunspace and living space, it should be 12 to 16 inches thick, and uninsulated.

A wall made of water containers 6 to 12 inches thick also works well. Water can hold a great deal of heat and deliver it easily to both rooms through convection currents in the water itself. Vents or windows and doors will help transfer warm air to the living space earlier in the day.

Wall openings that allow warm air to pass through should be included no matter what type of common wall is used. Warm air can flow from sunspace to living space through the top of a doorway or window and cool air can return through the bottom. This can also be done with pairs of high and low vents. There should be at least 8 square feet of openings for each 100 square feet of south glazing. When they are closed, the sunspace will act as a buffer against living area heat loss.

Any outside surface not used for collection or storage should be well insulated. The side walls of a sunspace should only be glazed if it is to be used for serious plant growing. East-west exposure supplies a negligible amount of solar heat. If the layout of the house permits, it can "wrap around" the sides of the sunspace, thus partly enclosing it. This design reduces heat loss, offers a place for more thermal mass than the add-on sunspace, and transfers heat to a larger area of the house itself.

As in any energy-efficient construction, care must be taken in putting together a sunspace to limit air infiltration. Tight construction, weatherstripping, and careful caulking to seal cracks are all important, especially around windows and vents that open and close.

Shades and vents are absolutely necessary for summer comfort. A properly-sized fixed overhang on vertical south glazing can be sufficient to keep the sunspace tolerably cool. If the glazing is tilted, a fixed overhang is impossible, and exterior shading with vegetation and interior shading with movable insulation or window shades are the only options.

Warm air in the winter carries desirable heat through doors and windows in the common wall to living rooms inside the house. But in the summer, that warm air must be vented outside. This is usually done with vent openings placed high and low on exterior walls of the sunspace. By convection, hot air rises out the top and draws cooler air in through the bottom.

Fans and thermostatic controllers may be necessary to move air, especially in the summer if sloped glazing is used. They can also be used to raise and lower movable insulation, to keep

Sunspaces are attractive additions to a home, for living space, winter greenery, extra heat, or a combination of all three. (Photo courtesy of Sunplace, Inc.)

temperatures more nearly constant automatically.

Sunspace Uses

The combination of design factors in a particular sunspace depends on its purpose. Is it to be a greenhouse? Is it purely a solar collector? Will it be used as a daytime and/or nighttime living space? Or will it be a combination of all three? Each type has its own design requirements.

A sunspace for serious year-round plant growing needs light as much as heat, so it should be glazed all around. Extreme temperatures are hard on plants, so thermal mass should be included to moderate them. Movable insulation or high-performance glazing are needed for night protection. Ventilation (both summer and winter) is also necessary to promote growth and control humidity.

If you opt for a solar living room, night heat losses are high, so vertical glazing insulated with movable insulation or special glazings is called for. To save heat, the east and west walls should be well insulated or enclosed by the house itself.

A sunspace used purely as a solar collector should have this same end-wall protection, because east-west glass gains little. Close it off at night and you can eliminate the cost of movable insulation and thermal mass.

The combination sunspace is the most popular—even though it requires some sacrifice in the efficiency of any one purpose. Most people want the space to serve many purposes. It should have as many of the features described above as possible, but compromises will be necessary to balance the various requirements of comfort during the day and night, winter and summer.

PASSIVE VERSUS ACTIVE SYSTEMS

The real beauty of passive solar design is its ability to function without external power sources. But as the three functions of solar heating—collection, storage, and distribution—become more distinct, external mechanical power is needed to transport the heat. Natural air flow on a large scale requires very large ducts that are too expensive, so a pump or fan is necessary. But how do you know which system will perform best in your particular situation?

The advent of hand-held calculators and microcomputers has made three major solar calculations generally usable. For passive solar calculations, Los Alamos National Laboratory's Load Collector Ratio (LCR) and Solar Load Ratio (SLR) are the most popular. Each predicts the performance of direct gain systems, mass walls, and sunspaces. The LCR method helps you determine *annual* performance in the early stages of designing. The LCR is the building's heat loss per degree day divided by the area of south glazing.

The SLR method determines performance on a *monthly* basis. Its more detailed outputs are useful later in the design process. The SLR is the ratio of solar gain to heat load, and is used to find the Solar Savings Fraction (SSF)—the ratio of the energy savings from solar to the net heating load of the same buiding without solar heat. Outputs include monthly and annual SSF, annual auxiliary energy use, and life-cycle solar savings.

F-Chart, from the University of Wisconsin, estimates the performance of passive and active solar energy systems for space heating, swimming pool heating, and domestic hot water. With this interactive program, you can analyze air or liquid systems, passive direct gain and mass walls, and swimming pool heating systems. Typical outputs include the solar gain, load, and fraction of the load met by solar.

Most of the microcomputer programs for determining system performance include life-cycle costing that help determine the most cost-effective design for a particular application. They also usually include weather data for several hundred locations. Other kinds of interactive solar software are coming onto the market, including programs for evaluating photovoltaic systems, daylighting, shading, and even wind turbines.

III

Solar Domestic Hot Water

I believe the ground rules can be transformed so that technology simplifies life instead of continually complicating it.

Steve Baer

The use of solar energy to heat household water supplies has been technically feasible since the 1930s, when solar water heaters were commonly used in California and Florida. Solar domestic water heating made a comeback in the 1970s and continued to sell even when oil prices dropped in the 1980s and the sale of active solar space conditioning systems plummeted. Their smaller scale and lower cost put them within closer reach of the homeowner's pocketbook. Furthermore, they integrate more easily with existing water heating systems.

Hot water needs are fairly constant throughout the year. The collector and other parts of the system operate year round, and initial costs can be recouped more quickly than with space conditioning systems. A solar space heater is fully operational only during the coldest months of the year, and the payback period is longer.

A solar water heater can also be sized more closely to the average demand. A water heater has roughly the same load day in and day out and doesn't have to accommodate wide fluctuations in demand.

A problem common to *all* types of solar heating is the fluctuation of available sunlight. But variable weather conditions are less problematic for household solar water heating because hot water requirements are more flexible. If the supply of hot water runs out during extended periods of cloudy weather, the consequences are less severe than if the house were to lose its heat. It's the difference between letting the laundry wait a bit longer or having the pipes freeze and burst. If you can tolerate occasional shortages of hot water, your solar water heater can have a very straightforward design—free of the complications that provide for sunless periods.

When a more constant hot water supply is needed the existing, conventional water heater can make up the difference. Controls are simple enough because this auxiliary heater can boost incoming water temperatures to the desired supply temperature. If the solar heater is providing full-temperature water, the auxiliary remains off. If not, the auxiliary comes on just long enough to raise the solar-heated water to the required temperature.

In addition to all these advantages, solar water heaters are a lot smaller than solar space heaters. The initial cost of a solar water heating system is lower, and it can be installed and operating within a very short time.

How large a system you need depends on how much hot water you use daily, and whether you will draw it directly from the solar heater, or use the solar heater to preheat the water for a final boost by a conventional water heater. On average, a solar water heater will supply 50 to 75 percent of your annual hot water needs; 25 to 35 percent in the winter, and 50 to 100 percent in the summer, depending on how much water you use and what system you choose.

Where should you locate your collectors? The first thing to consider is that the collector should face south or as nearly south as possible. Because domestic hot water is required year round in about the same daily amounts, the collector should be tilted for about the same solar gain in all seasons. Relatively constant daily insolation strikes a south-facing collector tilted at an angle equal to the local latitude. Steeper tilts (up to latitude + 10°) may be needed in areas with limited winter sunshine.

If the roof can support the collector, it is much less expensive to put it there than to build a separate structure. (Before installing the collectors on the roof, be sure the roof is sound—if it is an older house, it may need reinforcement.) True south orientation and latitude angle tilt may not be feasible on many houses. Fortunately, the loss in efficiency is relatively small (10 to 15 percent from ideal) if the roof faces within 25° of *true* south and its tilt is within 15° of the latitude angle. This efficiency reduction can be easily recouped by making a proportionately larger collector.

How much storage capacity should you have? Two days is generally the optimum. A larger tank will carry a household through longer cloudy periods (assuming enough collector area), but will require more time to reach the desired hot water temperature. Every attempt should be made to stratify the hot water in the storage tank above the cold. The hot water then goes directly to the load, and the cold water to the collectors, increasing efficiency and allowing the use of a larger tank. The larger tank costs more but a smaller one requires more frequent use of auxiliary heat. In general, the collector should be large enough to provide a single day's hot water needs under average conditions. Beyond this point, a larger collector experiences the law of diminishing returns.

After the system is installed, check regularly on the ouside racks, fasteners, mounts, and insulated piping. They should be repaired immediately if damaged by weather. Follow the manufacturer's maintenance instructions and schedules faithfully.

Just as in space heating, there are passive and active solar DHW systems. The passive group includes batch or integral collector storage (ICS), thermosiphoning, and phase-change systems. The active group—those powered by an electrically-driven pump—include antifreeze, drainback, draindown, and recirculation.

A third group, photovoltaic-powered systems, can be totally passive because the sun provides the electricity to power the pump. But the system uses all the same components as an active system with the addition of the photovoltaics panel.

8
Passive Solar DHW Systems

Passive solar domestic hot water heaters come in a wide range of shapes, sizes, efficiencies, and costs. Many can be totally passive by using the sun's power in the collection mode and city water pressure in the storage and distribution modes. Many are termed "passive" but are actually hybrid systems, because although their collection modes may be passive, their storage and distribution loops require auxiliary power for pumping.

BATCH HEATERS

In the batch heater, the solar storage tank itself is the collector and all the water is heated together in one "batch." One or more pressure tanks are painted black and placed inside a glazed, insulated box. House water pressure draws the hot water from the tank and takes it directly to the user, or to the conventional water heater.

The batch design is the easiest to do yourself. The inside tank of a conventional water heater, minus the metal cover and insulation, do nicely. You can even hang one or more tanks in a sunspace or greenhouse.

Batch systems are the least efficient solar DHW systems, but they are also the least expensive. If it is going to be used year round in a cool climate, all the exposed pipes require heavy insulation, and double or triple glazing is recommended for the box. If installed in a freezing climate, the batch heater is usually drained before the first freeze and lies dormant for the winter.

The "bread box" is a batch water heater that retains the virtues of low cost and simplicity of design. Variations of this water heater were first used in the 1930s.

The bread box minimizes heat loss from the stored hot water by means of an insulated box that encloses the tank at night and during cloudy weather. During the day, a top panel is raised and the front panel on the south face is lowered to expose the glass and tank to sunlight. The inside surfaces of the panels and the box itself are covered with a material such as aluminum foil to increase collection by reflection of sunlight around to the sides and back of the tank. When the panels are closed these surfaces reflect thermal radiation back to the tank.

The tank can be filled with water from either a pressurized or non-pressurized source. Once in the tank, the water is heated slowly but uniformly. Convection currents and conduction through the tank metal distribute the heat throughout the water, and little heat stratification occurs. In unpressurized systems, the hot water is nearly used up before the tank is refilled. In pressurized systems, cold replacement

The Bread Box: a simple and effective storage-type water heater.

water is drawn into the tank as hot water is drawn off and some mixing occurs. If dual tanks are used, the bread box has less difficulty with mixing. Hot water is drawn from one tank while cold water flows into the other.

Freezing in cold weather is not much of a problem becuause of the large volume of warm water in the tank. However, the pipes must be protected with insulation and electric heat tape (if the system isn't drained for the winter).

Integral collector storage (ICS) systems are the modern, manufactured version of the bread box. The units are better insulated, and many manufactureres claim that if they are plumbed with polybutylene pipe, they can withstand multiple freezes. However, many local plumbing codes still do not allow the use of polybutylene pipe in potable water lines.

Batch and ICS heaters are really DHW ''pre-heaters,'' since their lower efficiencies rarely let them achieve the temperatures needed by the average family. Their efficiency is hurt drastically by the high heat loss from the tank to the ambient air at night—a loss other DHW tanks located inside the house don't experience. But if the demand for hot water is concentrated into the early evening when the water is its hottest and before outside temperatures drop, their maximum efficiencies can be reached.

ICS manufacturers are striving to make their units more attractive by lowering the profile of the collectors to look more like their ''flat-plate'' collector counterparts. One manufacturer does this by having many smaller stainless steel tanks plumbed in series in one collector instead of one or two larger diameter tanks. The smaller-profiled collectors can be roof mounted like other collectors, but the concentrated dead load of the 30 to 40 gallons of water on the roof, and the nighttime heat loss, still remain.

Another manufacturer solves all three problems by using a phase-change material in the collector instead of water. Copper pipe in fin tubes runs through long rectangles of wax. The wax melts during the day, storing latent and sensible heat (see the discussion of phase-change collectors below). It transfers the heat to the water running through the pipes when there is a demand. As it cools, the wax solidifies from the outside in, insulating itself against high evening heat losses. The collector only holds 5 gallons of water at one time, and is only 5.5 inches high.

THERMOSIPHONING WATER HEATERS

The least complicated type of flat-plate solar collector is one that thermosiphons. It has no pumps, controllers, or other moving parts in the collection loop. All that moves is the water. It operates on the principle of natural convection: the hot water rises from the collector to a tank located above the top of the collector.

The older thermosiphoning water heater designs and site-built systems have a completely separate collector and storage tank. Insulated pipes connect a tilted flat-plate collector with a well-insulated tank. In an open-loop system, the water in the collector is heated by the sun, rises, enters a pipe, and flows into the top of the storage tank. Simultaneously the cooler water at the bottom of the storage tank flows through another pipe leading down to the bottom of the collector. As long as the sun shines, the water circulates and becomes warmer.

Collector backs, pipes, and tank should all be insulated, with the insulation around the tank as thick as your pocketbook permits. Four inches of fiberglass isn't excessive. If possible, place the tank indoors—in the attic if the collector is on the roof, or in the room behind the collector

if the collector is near the south wall. That way, the heat lost will flow to the rooms. If the tank has to be outside, it should be shielded from the winds and lavishly insulated.

Manufactured thermosiphoning systems are installed as one unit, with collector and tank together. They are available in the open-loop systems described above, where water is used for collection and storage, or in closed-loop systems.

In a closed-loop system, the heat transfer fluid follows the same pattern, but passes through a heat exchanger in the storage tank. The heat transfer fluid is usually a mixture of glycol and water or other non-freezing mixture. This protects the collector from freezing, and a well-insulated tank protects the water within it from freezing. But the piping from the cold water supply and to the auxiliary heater in the house must still be protected.

To eliminate the large heat loss from the tank above the collector, many manufacturers have *replaced* the tank with a heat exchanger. The heat exchanger is connected to the storage tank, or between the supply line and the storage tank. If the collector heat exchanger (or tank, as in the above case) is plumbed between a pressurized supply line and the storage tank, it is totally passive. If the heat exchanger (or tank) is connected only to the storage tank, then the storage loop must be pumped, and the system is considered hybrid. When the heat exchanger above the collector is connected to another heat exchanger in the remote storage tank, and an antifreeze solution is used, the system is completely protected against freezing.

In addition to lowering the heat loss from the tank, the use of a heat exchanger instead of a bulky tank above the collectors, lowers the profile of the collectors. It also eliminates the dead load of more than 40 gallons of water on the roof.

For thermosiphoning solar water heaters, the collector location must allow placement of the storage tank at a higher level. For roof-mounted collectors, the storage tank may be placed under the roof ridge or even in a false chimney. The

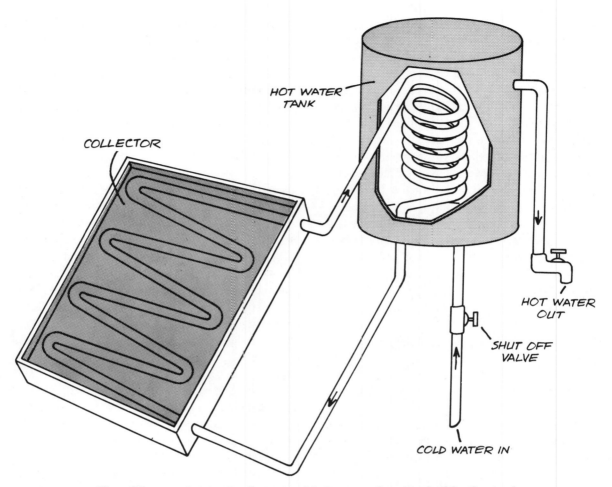

HOT WATER
TANK

COLLECTOR

HOT WATER
OUT

SHUT OFF
VALVE

COLD WATER IN

Closed-loop solar water heater with heat exchanger inside the tank.

roof structure must be strong enough to support the weight of a full tank. One alternative is to build a collector support structure on the ground, detached from the house or leaning against it. The collector then feeds an elevated tank located beside it or inside the house. Another possible location is on the roof of a lean-to greenhouse built onto the south-facing side of a house.

PHASE-CHANGE SYSTEMS

Materials can store two different kinds of heat, latent and sensible. *Latent* heat is stored when a material changes *phase* from a solid to a liquid and released when the material changes back to

a solid. Other phase-change materials change from a liquid to a gas and back. Latent heat is stored at one particular temperature. For example, when one pound of ice at 32°F melts, it absorbs 144 Btu, but the cold water's temperature is also 32°F. When it freezes, it gives up 144 Btu, but the temperature of the ice is still 32°F. *Sensible* heat is stored when a material is heated and rises in temperature. A single Btu is stored when you raise one pound of water 1°F. Phase-change materials, such as Freon, eutectic salts, or wax, change phase in a temperature range better suited to solar energy systems than water's 32°F freezing point and 212°F boiling point. For example, some salts melt at 84°F and store 75 Btu/1b—perfect for passive systems. Others melt at 97°F, and store 114 Btu—perfect for low-temperature active systems. (The pros and cons of phase-change materials are discussed in Chapter 15.)

Traditional phase-change collectors use a refrigerant that changes from a liquid to a gas when heated by the sun. The gas bubbles rise in the collector, pass over the heat exchanger, and condense to a liquid again as they give up their latent heat. The condensed fluid flows by gravity to the bottom of the collector to begin the process again. Since the fluid is able to change phase and store latent heat, more energy is collected over the same temperature rise than by thermosiphoning collectors, which store only sensible heat. Manufacturers claim increased efficiencies of 30 to 40 percent.

Although the collectors cannot freeze, the potable water side of the heat exchanger can. And most phase-change systems still need to have their heat exchanger installed above the top of the collector. Another drawback to phase-change refrigerant systems is the need for silver-solder connections, which increases cost.

A new phase-change system has been developed that is totally passive and yet still allows the storage tank to be located one story below. An evacuated closed-loop system is filled with a water-alcohol heat transfer fluid that changes phase (boils) at a low temperature. Its change in phase drives the fluid through the loop, and doesn't transfer heat between collectors and storage (as in traditional phase-change collectors). The fluid boils, carrying a mixture of gas and liquid to a riser across the top of the collectors. This causes a pressure difference in the closed loop, which the system naturally tries to overcome. The liquid portion of the mixture flows down to the storage tank, gives up its heat in the heat exchanger, and is forced up again to the collectors by the difference in pressure. Meanwhile, the gas in the header travels to the vapor condenser, where it condenses against the cooler liquid returning from the heat exchanger. The combined liquid falls through a pipe to the bottom of the collectors.

The wax ICS unit described in the section on integral collector storage systems is another type of phase-change system. But this time, the phase-change material is the collector's built-in storage and not the heat transfer fluid. The collector is plumbed between the cold water supply line and the auxiliary DHW tank. As hot water is drawn from the water heater, it is replaced by warm water from the collectors. The collector's absorber is made of wax, encased in long extruded-aluminum canisters coated with a selective surface. Copper-fin tubes, which help conduct the heat from the wax to the passing water in the tubes, are embedded in the wax. The wax holds onto its heat longer into the evening than traditional flat-plate collectors can. And it has a lower heat loss than other ICS units because the wax insulates itself as it cools and hardens from the outside in.

FREEZE PROTECTION

When air temperatures drop below 32°F, freezing water can burst the pipes or collector channels of a solar water heater. Less obvious but more dangerous is freezing caused by radiation to the night sky. Copper pipes in collectors have frozen and burst on clear, windless nights when air temperatures never dipped to freezing. The heat lost by thermal radiation was greater than that gained from the surrounding air.

To protect a passive solar DHW system that uses potable water in locations where freezing temperatures occur only now and then, heat tape fastened to the back of the absorber is a simple and inexpensive safeguard. Heat tape, commonly used to prevent ice dams on eaves, looks like an extension cord and has a small resistance to electricity. A thermostat turns it on when the outside temperature falls to 35°F, and the current flowing through it heats the absorber. In more severe climates where it would be called on frequently, heat tape would be too hard on the electric bill, and the batch heater, ICS, or open-loop thermosiphoning system would have to be drained for the winter.

Active Solar DHW Systems

The use of pumps can remove many of the architectural constraints of thermosiphoning water heaters. A pumped system is commonly used when piping runs would be too long or an elevated tank is impossible. The penalty paid for choosing an active system is the additional first costs of the pump and controls, and the electricity needed to run them. On the positive side, you have more freedom in the system layout. A pumped system can have a collector on the roof and the storage tank located anywhere you like. The best advantage is the additional useful energy produced per square foot of collector area.

Active solar domestic hot water systems prevent freeze-ups in one of three ways: by using an antifreeze solution; by draining the collectors; or by circulating warm water through the collectors. Basically, systems are divided into open- and closed-loops. Open-loops circulate potable water through the collectors. The collector array is pumped directly to the tank. Hot water is drawn from the top of the tank, and cold water from the supply line replaces it at the bottom. Cool water from the bottom of the tank is pumped to the collectors and solar-heated water is returned through a dip tube to the middle of the tank, as shown in the figure.

Two major problems can plague open-loop systems: corrosion and freezing. If water quality is poor, or freezing is a common winter occurrence, you'd be better off with a closed-loop system.

Closed-loop systems have separate collector and storage loops that pass through a heat exchanger. Each loop may have its own pump. The heat transfer fluid in the collector loop can be distilled water, treated water (to reduce corrosion), or an antifreeze solution.

Active solar DHW systems are further classified by their specific method of freeze protection. Open-loops include "recirculation" and "draindown" systems. Closed-loops include "drainback" and "antifreeze."

RECIRCULATION

Recirculation systems are only recommended for areas where freezing temperatures occur less than 20 days a year. When a freeze snap switch on the collector header senses the temperature has dropped to 40°F, it "snaps" shut and sends a signal through the controller to the pump. The pump circulates warm water from the storage tank through the collectors until the temperature of the snap switch rises over 50°F, when it opens and signals the pump to turn off.

There are two major disadvantages to this type of system. First, it's the energy collected

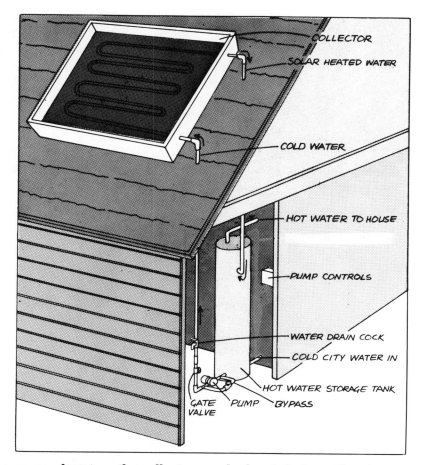

In a pumped system the collector can be located above the storage tank.

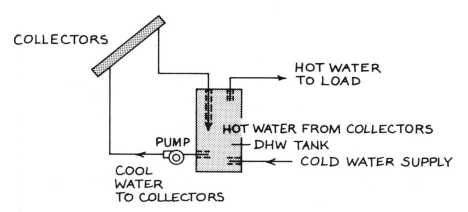

The Open-Loop System: Potable water from the tank also serves as the heat transfer fluid in the collectors.

77

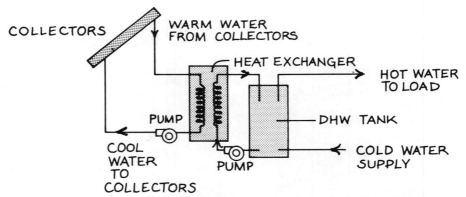

The Closed-Loop System: Treated water or antifreeze solution is circulated in the collection loop, which is separated from the storage loop by a heat exchanger.

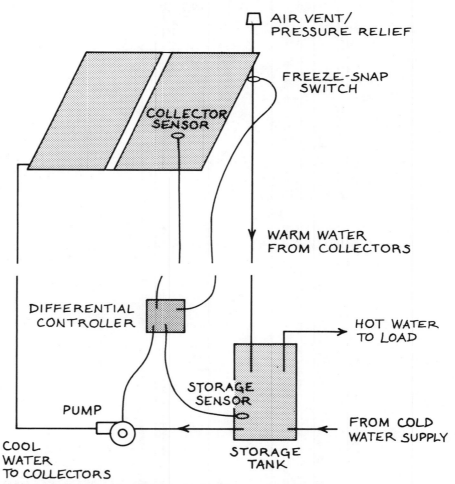

Recirculation System: When the freeze-snap switch senses the approach of freezing temperatures, it signals the controller to turn on the pump. Warm water from the storage tank circulates through the collectors until the snap switch temperature rises over 50°F, when it signals the controller to shut off the pump.

during the day that keeps the collectors and pipes from freezing at night. You don't want a recirculation system if that is going to happen often. Second, the freeze snap switch is electrically powered. If the freeze is accompanied by power outage, the snap switch won't be able to activate the controller and freeze damage could occur. Its one good point is that it uses water —the heat transfer fluid with the greatest heat storage capacity.

Like most active systems, the recirculation system uses a differential controller, which senses the temperature difference between the collectors and storage. The controller typically turns the pump on when the collectors are 10 to 15°F warmer than the storage tank, and off when they drop to only 2 to 3°F warmer than the tank. (You can buy a very accurate—but more expensive—differential controller with a 5°F-on and 1°F-off control strategy. This would allow the system to collect more energy.) The collector sensor is mounted on the absorber plate or collector header, and the storage sensor is mounted on the side of the tank under the insulation or on the supply line to the collectors where it exits the tank.

A recirculation design for pressurized systems relies on thermally-activated valves and city or well-pump water pressure to protect it from freezing. The freeze-protection valves, sometimes called dribble valves or bleed valves, use Freon or wax to change phase and open the valve at just above 40°F. The valve is placed on the return pipe leaving the last collector. Warm water from the pressurized tank flows through the collectors and spills out the valve onto the roof. Once the valve's temperature rises to above 50°F, the Freon or wax changes phase again and closes the valve.

The advantage is that the freeze protection doesn't rely at all on electricity. But the system still dumps heat from the storage tank to warm the collectors. Two to three gallons of warm water can pass through the valve before it closes. And if the valve sticks—open or closed— it could mean a lot of water lost down the roof, or damaged collectors.

DRAINDOWN

The first type of draindown system is basically the recirculation system with the addition of an electrically powered draindown valve. The valve is normally closed, except when the electricity to it is cut off. When the temperatures drop to just above 40°F, the snap switch signals the controller to cut the power to the valve. The valve automatically opens, and water drains out of the collector and supply/return piping. It is very important in systems that drain for freeze protection to pitch the collectors and pipes to drain *completely*. If they don't, trapped water could freeze and burst a collector riser or a pipe.

In pressurized draindown systems, another valve that is normally open is installed between the tank and collectors. When the snap switch closes, it signals the controller to also cut the power to the second valve so that it closes and prevents city water from refilling the collectors. When the snap switch opens, so does this second valve, and the collectors refill.

In either case, you must completely refill the system, and the pump must be large enough to overcome the pressure in the loop. An air vent/ vacuum breaker, located at the highest point of the collector array, opens to purge or draw in air when the system fills or drains. Unfortunately, the major problem with draindown systems is that the air vent can freeze shut, preventing the collector loop from draining.

An extra advantage of the draindown system is that since there is no fluid in the collectors, the system doesn't have to wait as long in the morning for the collectors to warm up before the controller can signal the pump to start.

DRAINBACK

Drainback systems have two separate loops for collection and storage with a heat exchanger between them. The collector loop is filled with a small amount of water, which is either distilled or treated to prevent corrosion. The sytem depends on a differential controller to activate the pump and fill the collector from a small tank

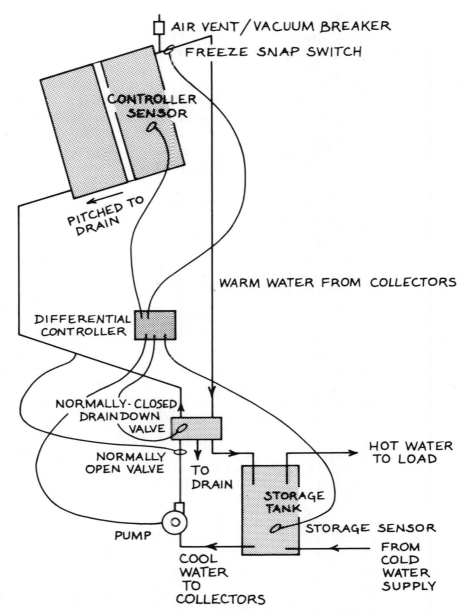

AIR VENT/VACUUM BREAKER

FREEZE SNAP SWITCH

CONTROLLER SENSOR

PITCHED TO DRAIN

WARM WATER FROM COLLECTORS

DIFFERENTIAL CONTROLLER

NORMALLY-CLOSED DRAINDOWN VALVE

NORMALLY OPEN VALVE

TO DRAIN

HOT WATER TO LOAD

STORAGE TANK

STORAGE SENSOR

PUMP

COOL WATER TO COLLECTORS

FROM COLD WATER SUPPLY

Draindown System: When temperatures approach 40°F, the freeze-snap switch signals the controller to open the draindown valve.

(8 to 12 gallons) that holds the heat transfer fluid. When the controller shuts off the pump at the end of the collection day, or when the collector temperature approaches 40°F, the fluid drains by gravity back into the tank. Just as in the draindown sytem, you have to be sure the collectors and piping are pitched properly to drain.

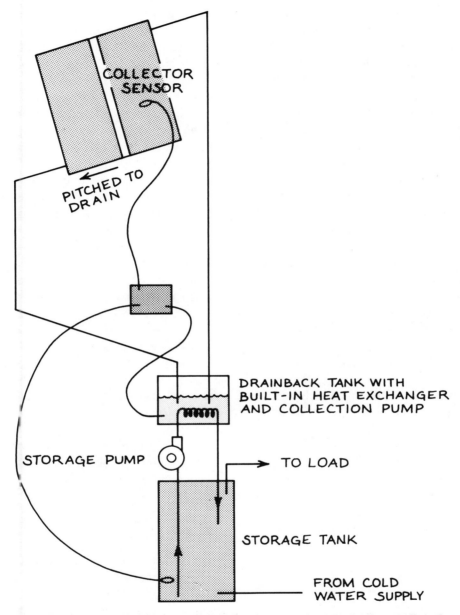

Drainback System: When the collector sensor signals the controller that temperatures are approaching freezing, it activates the drainback module to open its valves to drain the collector and return/supply piping.

Several manufacturers make drainback modules that include the insulated tank, heat exchanger, pump, and differential controller. One module has an air-vent/pressure-relief valve built into it that allows air to enter and escape when the system drains and fills. The tank completely fills with water when the system drains. When the pump fills the system, it creates a siphon that helps pull the water through the loop, so you can use a smaller pump. Unfortunately, the entry of outside air can mean increased corrosion in the collectors, piping, or tank. Another manufacturer's module has a tank with enough room for both air and water so that the system can be completely closed to outside air. This reduces corrosion, but it means you'll need more pump horsepower to overcome the pressure in the loop.

If the water in the tank is distilled or treated with a potable non-toxic corrosion inhibitor, a single-walled heat exchanger can be used. But if a toxic solution is used, the heat exchanger must be double-walled to prevent leakage into the potable water supply. Double-walled heat exchangers are less efficient than single-walled, reducing the total system efficiency. They're also more expensive.

Drainback systems require two pumps—one for collection and one for storage—unless the heat exchanger is part of the main storage tank. Tanks are available with heat exchangers inside them or wrapped around them, which transfer heat through natural convection currents in the water inside the tank. If the heat exchanger is separate from the tank, a circulator is needed to pump water to it from the tank.

The differential controller requires at least two sensors, with additional sensors and freeze snap switches recommended for extra freeze protection. Drainback systems are second only to antifreeze systems in popularity and freeze protection.

ANTIFREEZE

Antifreeze systems circulate a non-freezing heat transfer fluid through a closed collector loop. The collectors transfer their heat to storage through a heat exchanger. The primary advantage of these systems is that, depending on the heat transfer fluid, they can withstand freezing even in the most severe climates. In addition, a smaller circulator (a low horsepower pump) can be used, reducing first costs and annual

operating costs. The disadvantages are slightly lower efficiencies because of the less-effective heat transfer fluids and the heat exchangers they use.

There are many different types of heat transfer fluids for active systems. The most popular is a propylene glycol and water mixture. Propylene glycol is a non-toxic antifreeze, so a single-walled heat exchanger can be used. A double-walled heat exchanger must be used with ethylene glycol, its toxic cousin.

Other fluids include silicone, aromatic oils, paraffinic oils, and synthetic hydrocarbons. They each have their drawbacks, ranging from less desirable viscosities and lower flash points, to corrosion and toxicity.

Another non-freezing heat transfer fluid is air. It's non-corrosive, non-freezing or boiling, free, and it doesn't cause damage when it leaks. Unfortunately, its specific heat and density are much lower than the others, and coupled with an air-to-liquid heat exchanger, air solar DHW systems are much less efficient. The higher material and installation costs for ductwork over copper piping make them even less popular if you're only heating domestic hot water.

Just like drainback systems, antifreeze systems can have one pump or two, depending on the location of the heat exchanger.

PV-POWERED

Solar domestic hot water systems that include a photovoltaic (PV) panel are included in this section because they share many of the same components as active systems. The major difference is that the PV panel replaces the differential controller, power from the electric company, or both.

PV panels produce direct current (dc) electricity from the sun (see Chapter 16). The current can be used to signal the pump that there is enough solar insolation to begin collecting, or that there isn't enough to continue. In this case, the PV panel is only used to control, and you still purchase the electricity to run an al-

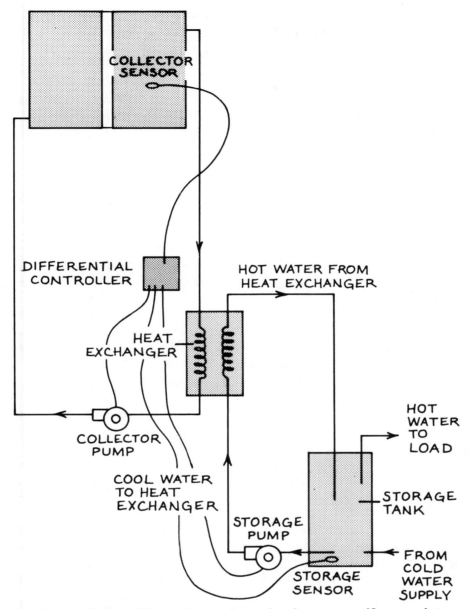

Antifreeze System: The collection loop circulates an antifreeze mixture through the collectors and the heat exchanger. The storage loop circulates potable water from the tank through the other side of the heat exchanger.

ternating current (ac) pump. If the electricity from the panel is also used to power a dc pump, then the system can be considered passive in nature.

Systems that rely on PV panels for control have their design problems. The insolation level at which the panel sends a signal to the con-

troller must be high enough to take into account the temperature of the absorber (which depends heavily on the ambient temperature) and the storage temperature (which depends on previous day's collection and water use). There could be little energy to collect if the absorber panel were slow to warm up because of sub-

freezing outdoor temperatures. Its shut-off insolation level must be set low enough to make
sure you can still collect the energy left in the
collector's materials themselves at the end of a
warm day. It also shouldn't allow the pump to
cycle on and off every time a cloud passes in
front of the sun. You have to carefully match
the PV panel to the pump to be sure it's big
enough to produce the extra burst of energy the
pump needs to start.

The more insolation available the warmer the
absorber plate, and the more energy being produced by the PV panel to power the pump. The
pump begins circulating slowly, increasing the
temperature of the water returned to the tank.
As the insolation level increases, the collector
absorber temperature increases and the PV panel
produces more electricity. The extra power makes
the pump circulate faster, increasing collector
efficiency by keeping the fluid temperatures
lower. Unfortunately, unless care is taken to
make sure tank water stays stratified in hot and
cold layers, the faster pump rate could stir up
the water in the tank and send warmer water to
the collectors—negating the increase in efficiency.

PV-powered controls and pumps are usually
designed for closed-loop, pressurized systems.
PV panels are expensive, and you need more
panels to produce enough power to overcome
the pressure in an open loop. A PV-pumping
system for an open loop can cost more than
three times as much as one for a closed loop.
And since the panels control the systems based
on changes in insolation and not temperature,
a separate freeze-protection mechanism must be
added.

ONE-TANK VS. TWO-TANK SYSTEMS

One-tank systems use the existing domestic water
heater for the storage tank and a backup heater.
If the tank uses electricity to heat the water,
remove the lower heating element so that heated
water isn't being pumped to the collectors. If
it's gas-fired, the tank must be specially man-

ufactured with the burner at the top. The cold
water supply line and the collector supply line
are connected to the bottom of the tank. The
collector return line and the line to the hot water
demand are connected at the top. as long as
water entering the tank moves slowly, the hot
water will stratify above the cold, so that only
cool water goes to the collectors. You can buy
special diffusers to slow the entry of water into
the tank. One-tank systems save you money
since you aren't buying an extra tank.

In two-tank systems, a tank for solar storage
is plumbed between the backup tank and the
collectors. the solar tank has no heating elements in it, but only stores solar energy. The
cold water supply line enters the bottom of the
solar storage tank. The top of the solar tank is
plumbed to the bottom of the backup heater,
which has a heating element or burner at the
top. As hot water is drawn from the top of the
backup tank, it is replaced from the top of the
solar-heated tank.

A two-tank system increases collector efficiency by returning cooler water to the collectors. Unfortunately, its standby losses are also
greater because of the increased storage volume. If the family is small, you may find the
single-tank system adequate, as long as you take
precautions to stratify the hot water above the
cold.

INSTALLATION CHECKLIST

It is very important that you follow the manufacturer's installation, operations, and maintenance instruction. But it will also be helpful to
ask yourself the following questions when designing and installing a solar domestic hot water
system:

• Are the collectors oriented properly? Do they
have an unobstructed view between 9 a.m. and
3 p.m.? Have you tilted them within acceptable
limits?
• Have you arranged the system components to
be easily accessible for service and repair?

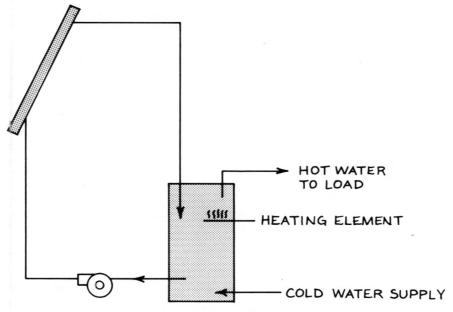

A one-tank solar DHW system.

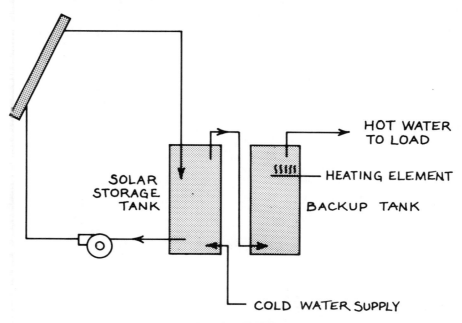

A two-tank solar DHW system.

• If you're planning on mounting the collectors on the ground, are they arranged so that they don't block drifting snow, leaves, and debris? If you're mounting the collectors on the roof, will the roof be able to support the additional load?

• Is the collector frame designed to support collectors under the most extreme local weather conditions? Can the frame material resist corrosion? Are the roof penetrations caulked or flashed to prevent water leakage? Have you installed the collectors so that water flowing off warm collector surfaces can't freeze in cold weather and damage the roof or wall? In areas that have snow loads over 20 pounds per square foot or greater, have you made sure that snow or ice sliding off won't endanger persons or property?

• Have you designed the system to follow the local and national codes that apply? Have you obtained the required building, plumbing, and electrical permits?

• Are all the pipes properly insulated to maintain system efficiency? Have you protected all exposed insulation from the weather and ultraviolet rays? Do you have enough pipe hangers, supports, and expansion devices to compensate for thermal expansion and contraction? If you're installing a draindown system, are the collectors and pipes properly pitched to drain all the fluid in areas where fluid might freeze?

• Have you designed in isolation valves so that major components of the system (pumps, heat exchangers, storage tank) can be serviced without system draindowns? Have suitable connections been supplied for filling, flushing, and draining? Do you have temperature and/or pressure relief valves to prevent system pressures from rising above working pressure and temperatures?

• Is the storage tank insulated well? Are the piping connections to the tank located to promote thermal stratification? Is the storage tank properly connected to the backup water heater?

• Are all system, subsystems, and components clearly labeled with appropriate flow direction, fill weight, pressure, temperature, and other information useful for servicing or routine maintenance? Have all outlets and faucets on nonpotable water lines been marked with a warning label?

• Are you sure you know how the system operates, including the proper start-up and shutdown procedures, operation of emergency shutdown devices, and the importance of routine maintenance? Does the owner's manual have all instructions in simple, clear language?

IV
Active Solar Systems

Who does not remember the interest with which when young he looked at shelving rocks, or any approach to a cave? It was the natural yearning of that portion of our most primitive ancestor which still survived in us. From the cave we have advanced to roofs of palm leaves, of bark and boughs, of linen woven and stretched, of grass and straw, of boards and shingles, of stones and tiles. At last, we know not what it is to live in the open air, and our lives are domestic in more senses than we think.

Henry David Thoreau,
Walden

Active solar heating and cooling systems use large expanses of tilted, glass-covered surfaces to collect solar energy and convert it to heat. A fluid—either air or a liquid—carries this heat through pipes or ducts to the living areas or to storage units. As opposed to the methods and systems discussed in previous chapters, active systems involve more complex and interdependent components. Their elaborate collectors, fluid transport systems, and heat storage containers require a network of controls, valves, pumps, fans, and heat exchangers. From a cost standpoint, solar space heating and cooling systems may be more appropriate for apartment buildings, schools, and office buildings than for single family dwellings. But coordination between the owners, architect, engineer and contractor can produce an active solar energy system appropriate to the residential scale. The major role in designing residential solar space heating and cooling systems is to "keep it simple."

10
Space Heating and Cooling

Simplicity is the watchword in the overall design of a solar energy system. It's tempting to design more and more complex systems—always trying to squeeze one more ounce of performance or a little more comfort out of them. But this added complexity usually means higher initial costs and greater operating and maintenance expenses. It's better to design a simple system that may require the inhabitants to toss a log or two in a fire every now and then.

If you're installing the system in a new house, design the house to incorporate passive solar systems to collect and store solar heat in the walls and floors. On a sunny winter day, enough solar energy streams through a hundred square feet of south-facing windows or skylights to keep a well-insulated house toasty warm long into the evening. And if the house has a concrete floor slab or masonry walls insulated on the outside, any excess heat can be stored for use later at night. The solar heat gathered in the active collectors can then be stored away until it is *really* needed rather than squandered heating the house during a cold, sunny day.

HEAT TRANSFER FLUIDS

When designing an active solar system, you must choose a fluid for transporting the heat.

There are usually two primary heat transport loops: one links the solar collector to the heat storage container: the other delivers the heat from storage to the house. Liquids or gases may be used as the heat transfer fluid in either loop. Liquids including water, ethylene glycol, and propylene glycol have predominated. Air is the only gas that has been used. The following criteria influence the selection of a heat transport fluid:

- Personal needs and comfort.
- Compatibility with the backup heating system.
- Compatibility with other mechanical devices.
- Climate (notably freezing).
- Relative cost (initial, operating, maintenance).
- Relative complexity.
- Long term reliability.

When personal comfort requires only space heating, forced-air systems are favored because of their relative simplicity and long lifetimes. When domestic hot water must also be provided, cold inlet water can be *preheated* before reaching the hot water heater where it is then raised to its final temperature. This preheating can be accomplished by passing the cold water supply through a heat exchanger in contact with

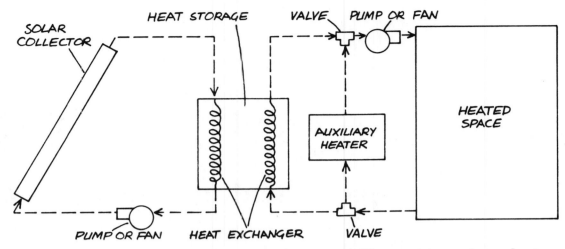

Basic components of an active solar heating system. There are two primary heat transport loops from collector to storage and from storage to the room.

the solar heated fluid in the return air duct to the storage bin or tank in an air system, or in the tank itself in a liquid system.

If cooling is needed in addition to heating, a liquid system is a more likely choice. Although some research has been done with air, most solar-powered cooling systems use liquids. The same thermodynamic and physical properties that favor liquids in conventional cooling units also favor them in solar cooling systems. Air systems, however, can be used successfully for some types of cooling. In arid parts of the country, for example, cool night air can be blown through a rock bed and the coolness stored for daytime use.

The method of distributing heat or coolness to the rooms may help you determine which fluids to use. Forced-air circulation is most compatible with air systems. Forced-air collectors store their heat in bins filled with rocks. When the house calls for heat, room air is blown through the rockbin to deliver the heat to the living spaces. But forced-air delivery systems can also be used with liquid solar collectors. Warm or cool water from the storage tank is passed through fan-coil units or heat exchangers, where the air blown across them is heated or cooled and delivered to the house. Because of the hot or cold drafts that occur, forced-air heating and cooling sys-

tems *can* be uncomfortable to the people using them and they must be designed carefully. But they do have the advantage of greater simplicity.

Most radiant heating systems use water to transport the heat, but some use hot air circulated through wall, ceiling, or floor panels. Hot water radiant systems, such as baseboard radiators, work well with liquid systems. Hot water from high-temperature collectors can circulate directly through a baseboard heating system or be sent to the heat storage tank. The main disadvantage is the high (140°F to 190°F) water temperatures. The higher the water temperature used, the lower the overall efficiency of the solar heating system. Steam heating systems are generally incompatible with solar collection because of the poor operating efficiencies of collectors at those high temperatures (with the exception of the more expensive concentrating and evacuated tube collectors). But many designers are installing liquid systems that circulate their fluid through polybutylene tubes in concrete floors. The concrete stores the heat and radiates it to the living space. These active-charge/passive-discharge systems are becoming very popular since they don't require the higher temperatures that baseboard heating systems demand.

COMPONENT OPTIONS FOR ACTIVE HEATING SYSTEMS

Collector Fluid	Heat Storage	Heat Distribution
Air	Building itself	Natural convection
	Rocks or gravel	Thermosiphoning
	Small containers	Forced convection of water
	Small containers of phase-change materials	Air fed radiant panels or concrete slabs
Water	Building itself	Natural convection
	Large tanks of water or other liquids	Baseboard radiators or fan-coil units
Water-antifreeze solutions	Polybutylene tubing in concrete slabs	Water-fed radiant panels or concrete slabs
Oil and other liquids or phase-change materials	Large tanks of water or other liquids	Forced convection past water-to-air heat exchangers or heat pumps

The amount of space allotted to heat storage is often a critical factor in the choice of fluids. Until phase-change materials are cheap and reliable, the main choices for heat storage are water and rock. Water tanks occupy from one third to one half the volume of rock beds for the same amount of heat storage. This fact alone may dictate the choice of a liquid system. The options available for collection, storage, and distribution of heat are summarized in the accompanying chart.

A choice of heat transfer fluids is available for residences—but not for larger buildings. The larger the solar-heated building, the greater the amount of heat that must travel long distances. If the fluid temperature is kept low to increase collector efficiency, either a heat pump is needed to raise the delivery fluid temperature, or more fluid must be circulated to provide enough heat to the building. Liquid heat delivery is better suited to large buildings because piping occupies less valuable space than ductwork. For air to do a comparable job, large ducts or rapid air velocities are necessary. Both alternatives are usually expensive, and the latter can be very uncomfortable.

Climate may dictate the choice of fluid. In cold climates, where a house may require only heating, air systems could be the most likely choice. When a liquid system is subject to freezing conditions, an antifreeze and water solution may be necessary. An alternative is to drain the water from the collector when the temperatures approach freezing.

First costs for materials and installation are also a factor. Storage and heat exchangers (or the lack of them) can cost less for air systems. Local labor economics often favor the installation of air ducts over water pipes. But don't underestimate the cost of fans and automatic dampers.

Air systems can be cheaper to maintain because air leaks are nowhere near as destructive as water leaks. Antifreeze solutions in liquid systems deteriorate and must be changed every two years. True enough, the cost of changing

91

antifreeze in cars and trucks is minimal, but a residential liquid-type solar heating system requires up to *50 times* as much antifreeze as a car. Air systems, on the other hand, can be more costly to operate than liquid systems because more electrical power is required to move heat with air than with water.

In all fluid transport systems, the network of ducts and piping should be kept simple. Pipes or ducts should be well insulated and as short as possible.

AIR SYSTEM DESIGNS

The very simplest active solar heating system has collectors that function only when the sun is shining and the house needs heat. Air is ducted

from the house to the collector, heated by the sun, and fan-forced into the room. The only heat storage container is the fabric of the house itself—and the heavier it is, the better. The fan operates when the collector temperature is warmer than that inside the house. It shuts off when the collector cools in the late afternoon or when room temperatures become unbearably hot. The more massive the house, the more heat it can store before temperatures get out of hand, and the longer it can go without backup heating. This type of system eliminates the controls, ductwork, and storage unit of the more expensive systems, and is becoming very popular.

Another simple active system delivers solar heated air to a shallow rock storage bin just beneath the house floor. Heat can flow up to

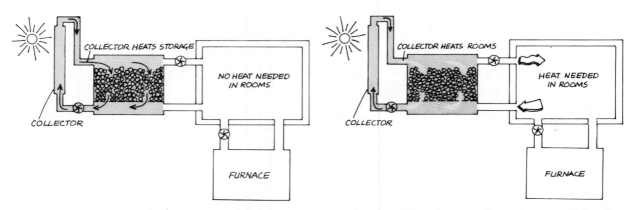

When the sun shines, the collector heats the storage. Storage is bypassed if the rooms call for heat.

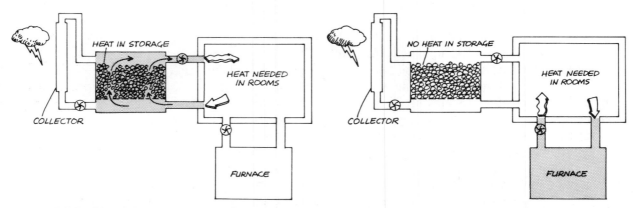

When the sun isn't shining, stored heat is delivered to the rooms as needed. If there is no heat in storage, the furnace comes on.

the rooms by natural convection through grilles, or if the rockbin is below a concrete slab, the heat is conducted through the floor and radiates to the space.

More storage for longer periods without sun is possible with another fan added to blow solar heat from a storage bin to the rooms. The fan draws cool room air through the storage bin and blows warm air back to the rooms. The backup heater (which can be a wood stove, electric heater, or an oil or gas furnace) can be in line with the solar storage, or be completely independent of the solar heating system. Ideally heat from storage isn't needed when the sun is shining because the solar heat gain through windows keeps the house warm. The heat from the collectors can be stored for later use. But if the house calls for heat, the collectors bypass the storage loop and supply it directly to the house. When the sun isn't shining and the house needs heat, solar heat is drawn from the storage bin —if available. If not, the backup heater is put to use. There are four possible modes of operation for this system and they are detailed in the diagrams.

In larger houses, it's expensive to have separate delivery systems for solar and auxiliary heat. Intergrating the two into a single delivery system requires extra dampers and controls but can be cheaper in the long run since no new ductwork is added. In any air system, extra ducting and dampers should be kept to a minimum.

LIQUID SYSTEM DESIGNS

Usually a liquid solar energy system is not as economical as an air system for heating a single-family dwelling. But with larger dwellings and increasing needs for domestic water heating and absorption cooling, a liquid system becomes more feasible. It doesn't have to be elaborate. A very basic liquid system for non-freezing climates is illustrated in the first of the two accompanying diagrams. Water from the storage

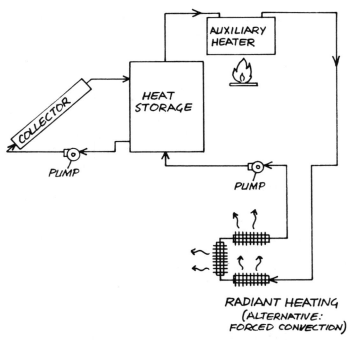

Piping system design for a simple liquid system.

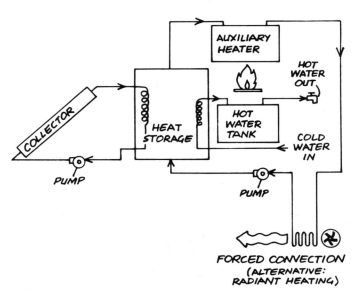

A liquid system designed for forced-convection heating and preheating of domestic hot water.

tank is heated by the auxiliary, if necessary, before delivery to the baseboard radiators or radiant heating panels. Only two pumps are needed to circulate the water through the two heat transport loops.

A somewhat more complex system is illustrated in the second diagram. Because of the threat of freezing, a water-glycol solution circulates through the collector and surrenders its solar heat to a heat exchanger immersed in the storage tank. Heat is distributed to the rooms by a warm air heating system that uses a fan to blow cool room air past a water-to-air heat exchanger. Cold inlet water from a city main or well pump passes through yet another heat exchanger immersed in the storage tank. This water is preheated before it travels to the conventional hot water heater. This type of system can be sized only slightly larger than a solar DHW system for each fan-coil unit connected to it.

An active solar energy system can be much more involved than these simplified diagrams indicate. Additional pipes, controls, and valves are required for the various modes of operation. Each heat exchanger degrades the overall performance of the system. The use of a heat exchanger substantially increases the collector operating temperature and lowers its efficiency. The greater the number of heat exchangers in a system, the lower the collector efficiency. Even more pumps, fans, and heat exchangers are needed than shown in the diagrams if solar absorption cooling is desired.

SWIMMING POOL HEATING

Solar collectors can be used to heat swimming pool water too. For outdoor swimming pools, inexpensive unglazed collectors can extend the swimming season earlier into the spring and later in the fall. How much longer you'll be able to swim depends on how cold it is in your area. Glazed collectors can be used in freezing climates to heat water year-round for indoor swimming pools.

Unglazed plastic or metal collectors perform well because they operate at low temperatures, usually in ranges that are only 10 to 20°F above the summer outdoor temperature. Since they don't need glass or plastic covers, they are less expensive than collectors that produce higher temperatures for space heating, cooling, and domestic hot water.

Because the pool water can be very corrosive when proper pH and chlorine levels aren't maintained, polybutylene or PVC plastic collectors are recommended more often for open-loop systems than copper collectors. They are more resistant to corrosion, but less resistant to ultraviolet radiation. Closed-loop systems—where treated water or antifreeze is spearated from the pool water by a heat exchanger—are recommended for metal collectors to avoid corrosion. But the closed-loop system can be less efficient and more expensive since it requires an extra pump and heat exchanger.

Most solar swimming pool heating systems have open loops and use the swimming pool's existing filtration pump. When the collectors are hot enough, the differential control signals a diverter valve to send the pool water through the collectors before returning to the pool. The diverter valve is located after the filter so that only clean water passes through the collectors. Closed-loop systems have an extra pump to circulate heat transfer fluid, and use the filtration pump to circulate pool water only.

PV-controlled systems are available for pool heating. At a preset time in the morning, pool water begins circulating through the filter. When the intensity of the sunlight reaches a preset level, the PV panel signals the controller to divert water through the collectors. When pool water reaches a preset temperature, the diverter valve bypasses the collectors and sends the water straight back to the pool. If pool water drops below the desired temperature, the diverter valve sends the water back through the collectors. At the end of every day, when the sunlight drops to a preset level, the valve diverts the pool water back to the pool filtration loop again. Finally at a preset time, the circulation pump turns off.

In hot-arid climates, the cycle can be reversed to cool the pool during the summer. Pool water is circulated through unglazed collectors at night to radiate the heat to the night sky. A

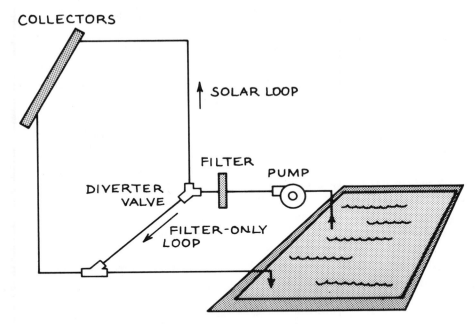

COLLECTORS

SOLAR LOOP

FILTER

PUMP

DIVERTER VALVE

FILTER-ONLY LOOP

Solar swimming pool heating.

timer turns the pump on at night and off in the morning for a more refreshing water temperature.

Another pool heating system has polybutylene pipes buried in a poured concrete slab around the pool. Pool water is circulated through the tubes, cooling the solar-heated patio as it warms the pool water.

How large a collector area is needed will depend on the kind of collector choosen. Unglazed plastic collectors usually have an area one-half to three-quarters the pool area. Glazed collector systems require much less: 40 to 50 percent of the pool area. Patio systems must be bigger because they are less efficient: about 130 percent of the pool area.

CONTROLS

One set of controls governs the delivery of heat (or coolness) to a house from the collector, heat storage, or backup heating (or cooling) system. Its operation is determined by the needs of the household and the limits of the entire system. In general, the thermostat governing the energy flow from storage can operate at a different temperature level than the thermostat on the backup heater. Often a two-stage thermostat is installed. The first setting might be at 70°F and the second at 68°F. If the heat storage cannot maintain 70°F room temperatures, the backup system springs into action when the temperature falls below 68°F.

Controls to govern collector operation are relatively simple and are readily available. Most of these controls determine collector operation by comparing the collector temperature and the storage temperature. One temperature sensor is placed directly on the absorber. The other sits in the storage tank or near the return pipe to the collectors. Customarily the collector pump starts working when the collector is 10 to 15°F warmer than the storage. For air systems, a temperature difference of as much as 20°F may be needed before the circulation fan is triggered. A time delay of about 5 minutes is necessary to prevent the system from turning on and off during intermittent sunshine. Some liquid systems may need controls that prevent liquid temperatures from rising to the point where pressures can cause piping to burst or degrade the heat transfer fluid.

Photovoltaic panels, that convert solar energy to electricity, are also being used to control and pump solar systems. The photovoltaic panel turns the pump on when solar insolation reaches a certain level and turns it off when it falls below another level.

PERFORMANCE AND COST

The tradeoff between performance and cost is crucial to the design of any solar energy system. The performance of a system is measured by the amount of energy it can save a household per year. The dollar value of the energy saved is then compared with the initial (and operating) costs of the system. The initial costs must not get so high that they can never be recouped over the life of the system. One doesn't have to be quite as careful in the design of conventional heating systems because the fuel costs are far and away the major heating expense. But the initial costs of an active solar heating system are usually so high that more than 10 years of trouble-free operation are needed before the energy savings make it a good investment.

SOLAR COOLING

Active solar energy systems can also cool a house during the summer. And the sun is usually shining the brightest when cooling is needed most. The hottest months and times of day occur at times of nearly peak solar radiation. Systems that provide both heating *and* cooling can operate the year round—with additional fuel savings and a shorter payback period.

Solar cooling seems paradoxical. How is it that a heat source can be used to *cool* a house? One answer is that solar energy is also a source of *power* that can move room air in ways that enhance comfort.

Substantial cooling can be obtained by using nocturnal radiation to cool the storage container at night. Warm objects radiate their heat to the cooler night sky—particularly in arid climates.

Warm air or water from storage is cooled as it circulates past a surface exposed to the night sky. The cooled fluid returns to the storage container, which is cooled in the process. The next day the storage is used to absorb heat from the house.

Solar collectors can provide the heat needed by an *absorption cooling* device—making solar-powered air conditioners a distinct possibility. An absorption cooling unit uses two working fluids—an *absorbent* such as water, and a *refrigerant* such as ammonia. Solar heat from the collector boils the refrigerant out of the less volatile absorbent. The refrigerant condenses and moves through a cooling coil inside the room. Here it vaporizes again, absorbing heat from the room air. The refrigerant vapor is then reabsorbed in the absorbent, releasing heat into cool water or the atmosphere.

Unfortunately, most absorption cooling devices work best with fluid temperatures between 250°F and 300°F. The lowest possible working fluid temperature that can be used is about 180°F—where flat-plate collectors have sharply reduced efficiencies. And the collectors have to operate at temperatures about 15°F to 20°F above this lower limit.

If 210°F water is supplied by a collector, the working fluids will receive solar heat at 180°F and the water will return to the collector at 200°F. On a hot summer day, a square foot of collector might deliver 900 Btu—or about 40 percent of the solar radiation hitting it. About 450 Btu will be removed from the interior air, so that a 600-square-foot collector can provide a daytime heat removal capacity of about 270,000 Btu or 30,000 Btu per hour.

Solar collectors designed for absorption cooling systems are more expensive than those used only for winter heating. But substantial fuel savings are possible if the same collector can be used for both purposes. Concentrating collectors and evacuated-tube collectors are particularly well suited to absorption cooling because they can supply high temperatures at relatively high efficiency. Almost all absorption cooling equipment requires liquid collectors.

Complete packages are available that combine solar space heating and domestic hot water, and evaporative and desiccant cooling. The cooling cycles are not solar, but first costs are lower for the whole system since it comes as a manufactured unit.

Absorption Cooling Principles

Just like window air conditioners and heat pumps, an absorption cooling device uses the evaporation of a fluid refrigerant to remove heat from the air or water being cooled. But window air conditioners and heat pumps use large quantities of electricity to compress this evaporated fluid so that it condenses and releases this heat to the "outside." The condensed fluid then returns to the evaporating coils for another cycle.

In an absorption cooling cycle, the evaporated refrigerant is absorbed in a second fluid called the "absorbent." The resulting solution is pumped to the "re-generator" by a low-power pump. Here, a source of heat—which can be fossil fuel or solar energy—distills the refrigerant from the solution.

The less volatile absorbent remains a liquid and returns to the absorber. The refrigerant liquid returns to the evaporating coils—where it evaporates and cools the room air, completing the cycle.

Absorption cooling devices can use hot fluid from a solar collector to boil the refrigerant from the absorbent. Unfortunately, most absorption cooling devices work best with fluid temperatures between 250°F and 300°F. Flat-plate collectors are inefficient at such high temperatures, but concentrating and evacuated-tube collectors can produce these temperatures easily. If their costs and complexity can be brought down, they may someday find an application in solar absorption cooling.

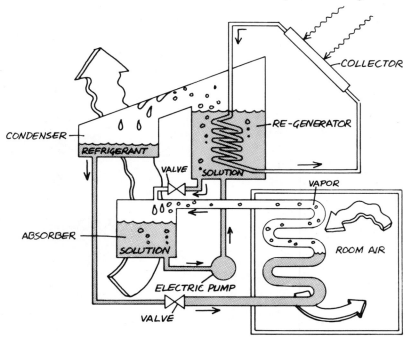

Absorption Cooling Cycle.

11
Liquid Flat-Plate Collectors

The primary component of an active system is the solar collector. It converts the sun's radiant energy into useful heat energy that is carried into the house by a fluid. The distinguishing feature of a flat-plate collector is that the sun's energy is absorbed on a flat surface. Flat-plate collectors fall into two catagories—*liquid or air*—according to the type of fluid which circulates through them to carry off the solar heat. A new circulatory fluid—phase change—falls into the liquid catagory, since it also circulates through tubes.

The basic components of a liquid flat plate collector are shown in the diagram. The absorber stops the sunlight, converts it to heat, and transfers this heat to the passing liquid. Usually the absorber surface is black to improve efficiency. To minimize heat loss out the front of the collector, one or two transparent cover plates are placed above the absorber. Heat loss out the back is reduced by insulation. All of these components are enclosed in a metal box for protection from wind and moisture. Most contractors will buy a manufactured collector, but they should look closely at what goes into them before they buy. The materials and design of a collector are crucial in determining its efficiency and how long it will last.

ABSORBER DESIGN

There are two types of absorber designs—each characterized by the method used to bring liquids in contact with the absorber plate. The first category includes open-faced sheets with the liquid flowing over the front surface. The Thomason absorber, with water flowing in the valleys of corrugated sheet metal, is a good example of these "trickle-type" collectors. The second more popular category uses tubes connected to a metal absorber plate. A variation on the tube-in-plate is the extruded plastic collector used in swimming pool heating.

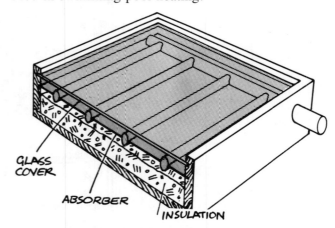

A liquid flat-plate collector.

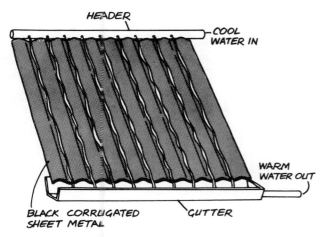

Thomason's trickle-down absorber.

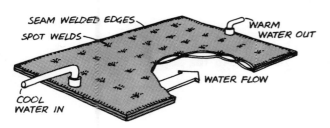

A typical sandwich-type absorber.

The open-face Thomason absorber shown in the diagram has the advantage of simplicity. Cool water from storage is pumped to a header pipe at the top of the collector and flows out into the corrugations through holes on top of each valley. A gutter at the base of the collector gathers the warm water and returns it to the storage tank. Its clearest advantage is that it is self-draining and needs no protection against corrosion or freezing. One disadvantage is that condensation can form on the underside of the cover plate. Another is that the trickling water may eventually erode the black paint.

In most of the early experimental work with flat-plate collectors, the absorber plates consisted of flat metal sheets with copper tubes soldered, welded, wired, or clamped to them. Thousands of experimenters all over the world have struggled to develop cheap, effective methods of bonding tubes to plates. Good thermal bonds are of paramount importance. Most commercially available collectors have copper tubes soldered to copper plates.

TUBE SIZING AND FLOW PATTERNS

The choice of tube size for an absorber involves tradeoffs between fluid flow rate, pressure drop, and cost. If cost were the only factor, the tube diameter would be as small as possible. But the smaller a tube, the faster a liquid must travel through it to carry off the same amount of heat. Corrosion increases with fluid velocity. And the faster the fluid flows, the higher the pumping costs.

Typically, the *risers* (the tubes soldered directly onto the absorber plates) are 1/2 inch in diameter, but this ultimately depends on the size of the system and the liquid being used. The *headers* (those tubes running along the top and bottom of the plate) are 3/4 to 1 inch in diameter.

The pattern of the tubes in the absorber plate is also important to the overall performance of the collector. Strive to attain uniform fluid flow, low pressure drops, ease of fabrication, and low cost. Uniform fluid flow is the most important of these. "Hot spots" on the absorber plate will lose more heat than the other areas—lowering overall efficiency.

Since most applications call for more than one collector, you will have to connect a number of independent collector panels together. Series or parallel networks are the simplest. Again, the important criteria are uniform fluid flow, low pressure drop, and the ability to fully drain the liquid in drainback systems. A network of collectors piped in series has uniform flow but a high pressure drop, while a parallel hookup has just the opposite. For a large num-

Tips on Corrosion Prevention

Because oxygen can be very corrosive under certain conditions, air should be prevented from entering the heat transfer liquid. This can be very difficult in self-draining systems.

The pH of the transfer liquid (a measure of its acidity) is the most critical determinant of corrosion. Liquids coming in contact with aluminum must be neutral—with a pH around 6 or 7. Any deviation, whether lower (more acidic) or higher (more basic) severely increases the

rate of corrosion. The pH must be measured frequently to prevent deviations from the norm. Antifreeze should be replaced at 12-month intervals.

Systems in which the transfer liquid·flows in contact with a number of different metals are susceptible to galvanic corrosion. If possible, you should avoid using several different metals. In particular, aluminum should be isolated from components made from other metals.

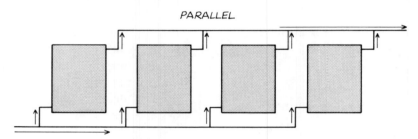

PARALLEL

Reverse return piping systems help balance the flow through the collectors. The first collector plumbed to the supply line is the last plumbed to the return line.

ber of independent collector panels, a series-parallel network is your best bet. In any network, the exterior piping should be at least 1 inch in diameter and well-insulated. Many collectors are available with integral top and bottom headers. Connections are made directly between collectors, reducing pipe costs and heat loss.

The plumbing configuration most often used is the reverse-return method, that follows the first-in, last-out rule. The first collector to receive liquid from storage is the last connected to the return to storage.

ABSORBER PLATES

Absorber plates are usually made of copper or aluminum. But plastics are taking over the low-temperature applications, such as swimming pool heating systems.

A metal need not be used for the absorber plate if the liquid comes in direct contact with

every surface struck by sunlight. With almost all liquid systems now in use, however, the liquid is channelled through or past the plate. Heat must be conducted to these channels from those parts of the absorber that are not touching the fluid. If the conductivity of the plate is not high enough, the temperatures of those parts will rise, and more heat will escape from the collector—lowering its efficiency. To reduce this heat loss, the absorber plate will have to be thicker or the channels more closely spaced. With a·metal of high conductivity such as copper, the plate can be thinner and the channels spaced further apart. To obtain similar performance, an aluminum plate would have to be twice as thick and a steel sheet nine times as thick as a copper sheet.

The accompanying graph illustrates the variation in absorber efficiency (the "efficiency factor" gauges the deviation from optimum) with tube spacing for various types and thicknesses of metals. Cost rises *faster* than efficiency for increasing thickness of copper.

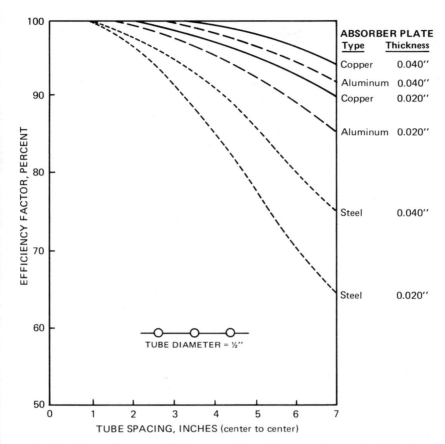

ABSORBER PLATE

Type	Thickness
Copper	0.040″
Aluminum	0.040″
Copper	0.020″
Aluminum	0.020″
Steel	0.040″
Steel	0.020″

TUBE DIAMETER = ½″

The variation in collector efficiency with tube spacing and absorber type.

Optimum cost and efficiency is achieved with a 0.010-inch-thick copper sheet with tubes spaced at intervals of 4 to 6 inches. Copper has become the most popular absorber choice in manufactured collectors.

ABSORBER COATINGS AND COVER PLATES

The primary function of the absorber surface or coating is to maximize the percentage of sunlight retained by the absorber plate. Any surface reflects and absorbs different amounts of the sunlight striking it. The percentage it absorbs is called its *absorptance* (α). *Emittance* (ϵ) is the tendency of a surface to emit longwave thermal radiation. An ideal absorber coating would have $\alpha = 1$ and $\epsilon = 0$, so that it could absorb all sunlight striking it and emit no thermal ra-.

diation. But there is no such substance, and we usually settle for flat black paints, with both α and ϵ close to 1.

There are a few substances called *selective surfaces* which have a high absorptance (greater than .95) and low emittance (less than .2). Selective surfaces absorb most of the incident sunlight but emit much less thermal radiation than ordinary black surfaces at the same temperature. Collectors with selective absorber surfaces attain higher collection efficiencies at higher temperatures than normal collectors. But they are necessary for systems which operate at temperatures below 100°F.

The absorber coating should be chosen together with the collector cover plate. They have similar functions—keeping the solar heat in—and complement each other in a well-designed collector. For example, a selective surface with

101

PROPERTIES OF SELECTIVE SURFACES FOR SOLAR ENERGY
APPLICATIONS

Surface	Absorptance for Solar Energy	Emittance for Long Wave Radiation
"Nickel Black" on polished Nickel	0.92	0.11
"Nickel Black" on galvanized Iron*	0.89	0.12
CuO on Nickel	0.81	0.17
Co_3O_4 on Silver	0.90	0.27
CuO on Aluminum	0.93	0.11
Ebanol C on Copper*	0.90	0.16
CuO on anodized Aluminum	0.85	0.11
PbS crystals on Aluminum	0.89	0.20

*Commercial processes. (Source: Duffie and Beckman, *Solar Energy Thermal Processes.*)

a single cover plate is usually more efficient than flat black paint with two cover plates. The accompanying graph compares the performance of flat black and selective surfaces for one and two cover plates. For collector temperatures below 150°F, a second cover plate may be superfluous, but for temperatures above 180°F (for process heat or absorption cooling) a second cover plate or a selective surface may be necessary. For temperatures below 100°F, a selective surface performs no better than flat black paint.

Cover plates are transparent sheets that sit about an inch above the absorber. Shortwave sunlight penetrates the cover plates and is converted to heat when it strikes the absorber. The cover plates retard the escape of heat. They absorb thermal radiation from the hot absorber, returning some of it to the collector, and create a dead air space to prevent convection currents from stealing heat. Commonly used transparent materials include glass, fiberglass-reinforced polyester, and thin plastics. They vary in their ability to transmit sunlight and trap thermal radiation. They also vary in weight, east of handling, durability, and cost.

Glass is clearly the favorite. It has very good solar transmittance and is fairly opaque to thermal radiation. Depending on the iron content of the glass, between 85 and 96 percent of the sunlight striking the surface of 1/8-inch sheet

of glass (at vertical incidence) is transmitted. It is stable at high temperatures and relatively scratch-and weather-resistant. Glass is readily available and installation techniques are familiar to most contractors.

High transmittance solar glass with a low iron content is used almost exclusively today in commercial solar collectors. Viewed on edge, the greener the glass, the higher the iron content and the lower the transmittance.

Alternatives to glass include plastic and fiberglass-reinforced polyester. Plastics, many of them lighter and stronger than glass, have a slightly higher solar transmittance because many are thin films. Unfortunately, plastics transmit some of the longwave radiation from the absorber plate. Longwave transmittance as high as 80 percent has been measured for some very thin films. The increased *solar* transmittance mitigates this effect somewhat—as does the use of a selective surface. But good thermal traps become very important at higher collector temperatures, and many plastics can't pass muster under these conditions.

Almost all plastics deteriorate after continued exposure to the ultraviolet rays of the sun. Thin films are particularly vulnerable to both sun and wind fatigue. Most are unsuitable for the outer cover but could be used for the inner glazing, with glass as the outer glazing. Some of the thicker plastics yellow and decline in solar

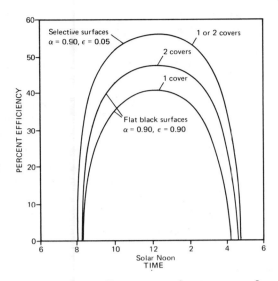

Flat-plate collector performance of selective and flat black absorber coatings.

transmittance, even though they remain structurally sound. Other plastics like Plexiglass™ and acrylics soften at high temperatures and remain permanently deformed. In dirty or dusty regions, the low scratch resistance of many plastics make them a poor choice. Hard, scratch-resistant coatings are available at an increased cost. Newer plastics are being introduced with special coatings to protect against ultraviolet degradation, with limited warranties up to 10 years against yellowing, scratching, and hail damage.

Additional cover plates provide extra barriers to retard the outward flow of heat and insure higher collector temperatures. Double-glazed commercially available collectors most commonly use two layers of glass. The more cover plates, the greater the fraction of sunlight absorbed and reflected by them—and the smaller the percentage of solar energy reaching the absorber surface.

In general, the lower the temperature required from the collector, the fewer the cover plates. For example, solar collectors that heat swimming pools usually don't require a cover plate. For cooler climates, additional cover plates may be needed. To obtain the same collector performance, for example, two covers may be

necessary in New England and only one in Florida. The majority of the collectors on the market have one layer of glass as the cover plate, and a selective surface.

INSULATION

Insulation is used behind the absorber to cut heat losses out the back. If the collector is integrated into the wall or roof, heat lost out the back is transferred directly into the house. This can be an advantage during winter but not in the summer. Except in areas with cool summer temperatures, the back of the absorber should be insulated to minimize this heat loss and raise collector efficiency. Six inches of high temperature fiberglass insulation or its equivalent is adequate for roof collectors, and as little as 4 inches is sufficient for vertical wall collectors if they are attached to a living space. Where the collector sits on its own support structure separate from the house an R-12 back and R-8 sides should be the minimum.

Choose an insulation made without a binder. The binder will vaporize at high temperatures and condense on the underside of the glazing when it cools, cutting transmittance. The insulation should be separated from the absorber plate by at least a 3/4-inch air gap and a layer of reflective foil. This foil reflects thermal radiation back to the absorber—thereby lowering the temperature of the insulation and increasing collector efficiency. Most collectors use a foil-faced cellular plastic insulation at the back of the absorber, separated from it with a layer of binderless fiberglass and a layer of foil. Both insulations are made *specifically* for high temperatures because the collector could stagnate above 300°F.

The perimeter of the absorber must also be insulated to reduce heat losses at the edges. Temperatures along the perimeter of the absorber are generally lower than those at the middle. So less insulation can be used, but it too should be made for high temperatures in case of stagnation.

The New Solar Home Book

OTHER FACTORS

Smaller issues should also be addressed when choosing a collector. Glazing supports and mullions can throw shade on the absorber so look for collectors with the standard low-profile aluminum extrusions. Gaskets and sealants should be able to resist ultraviolet radiation and high temperatures. The glazing details should provide for drainage and keep out snow, ice, water, and wind.

A filled collector weighs between 1 and 6 pounds per square foot. This is well below the roof design load of most houses. Wind loads on wall collectors or integral roof collectors are no problem either, since these surfaces must withstand wind conditions anyway. But, wind loads are important in the design of raised support structures for separated collectors.

Snow loads have not been a problem. The steep collector tilt angles needed at higher latitudes (where most of the snow falls) are usually adequate to maintain natural snow run-off. Even when snow remains on the collector, enough sunshine can pass through to warm the collector and eventually cause the snow to slide off. As a last resort, warm water from storage can be circulated through the collector in the morning.

12
Air Flat-Plate Collectors

Solar heating systems that use air as the heat transport medium should be considered for all space heating applications—particularly when absorption cooling and domestic water heating are not important. Air systems don't have the complications and the plumbing costs inherent in liquid systems. Nor are they plagued by freezing or corrosion problems.

The relative simplicity of air solar heating systems makes them very attractive to people wishing to build their own. But precise design of an air system *is* difficult. All but the simplest systems should be designed by someone skilled in mechanics and heat transfer calculations. Once built, however, air systems are easy to maintain or repair. Fans, damper motors, and controls may fail occasionally, but the collectors, heat storage, and ducting should last indefinitely.

The construction of an air collector is simple compared to the difficulty of plumbing a liquid collector and finding an absorber plate compatible with the heat transfer liquid. Except for Thomason's collector, the channels in a liquid collector absorber must be leakproof and pressure-tight and be faultlessly connected into a larger plumbing system at the building site. But the absorber plate for an air collector is usually a sheet of metal or other material with a rough surface. Air collectors must be built with an eye on air leakage and thermal expansion and contraction.

ABSORBERS

The absorber in an air collector doesn't even have to be metal. In most collector designs, the circulating air flows over virtually every surface heated by the sun. The solar heat doesn't have to be conducted from one part of the absorber to the flow channels—as in liquid collectors. Almost any surface heated by the sun will surrender its heat to the air blown over it.

This straightforward heat transfer mechanism opens up a wide variety of possible absorber surfaces: layers of black screening, sheets of glass painted black, metal lath, or blackened aluminum plates. Many of these can be obtained very cheaply—as recycled or reused materials. The entire absorber surface must be black, must be heated directly by the sun, and must come in contact with the air flowing through the collector.

A sheet metal absorber plate, the old standby for liquid collectors, is probably the best choice. Metal is preferable for collectors in which the sun cannot reach every last surface in contact with the moving air. Because of its high conductivity, metal can also alleviate the "hot spots" caused by an uneven air flow. Excess heat is conducted to other areas where the air is making better contact.

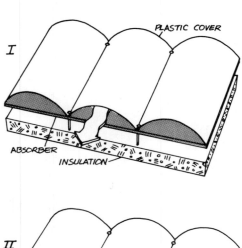

AIR FLOW AND HEAT TRANSFER

Just *where* to put the air passage relative to a blackened metal absorber is a question that merits some atttention. Three basic configurations are shown in the diagram. In Type I, air flows between a transparent cover and the absorber; in Type II, another air passage is located behind the absorber; and in Type III, only the passage behind the absorber is used. The Type II collector has the highest efficiency when the collector air temperature is only slightly above that outdoors. But as the collector temperature increases, or the ambient air temperature decreases, Type III is dramatically better because of the insulating dead air space between the cover and the absorber.

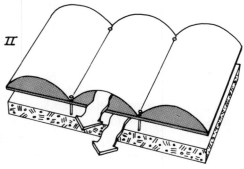

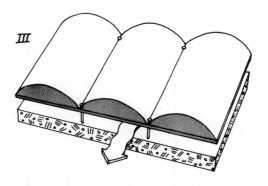

The rate of heat flow from the absorber to the passing air stream is also crucial. The *heat transfer* coefficient h is one measure of this flow. It is similar to the U-value, which is a measure of the heat flow through a wall or roof. The higher the value of h, the better the heat transfer to the air stream and the better the collector performance. Good values of h fall in the range of 6 to 12 Btu/(hr ft^2 °F). At a temperature 25°F above that of the air stream, one square foot of good absorber surface will transfer 150 to 300 Btu per hour to passing air—almost as much solar radiation as is hitting it. The value of h can be increased by increasing the rate of air flow, by increasing the effective surface area of the absorber, or by making the air flow more turbulent. As long as costs to run the fan or noise levels do not get out of hand, higher values of h are definitely preferred.

Whether the absorber surfaces are metal or not, *turbulent* flow of the air stream is very important. Poor heat transfer occurs if the air flows over the absorber surface in smooth, un-

The three types of warm air solar collectors.

disturbed layers. The air next to the surface is almost still and becomes quite hot, while layers of air flowing above it do not touch the absorber surface. Two levels of turbulent flow will help improve this situation. Turbulence on the macroscopic level can be observed with the naked eye when smoke blown through the air tumbles over itself. Turbulence on the microscopic level involves this tumbling right next to the absorber surface.

To create turbulent flow on either level, the absorber surface should be irregular—not smooth. Finned plate and "vee" corrugations create macroscopic turbulence by breaking up the air flow—forcing the air to move in and out, back and forth, up and down. To create microscopic turbulence, the surface should be rough or coarse, with as many fine, sharp edges as possible. Meshed surfaces and pierced metal plates do the trick.

But increased air turbulence means a greater pressure drop across the collector. Too many surfaces and too much restriction of air flow will require that a larger fan be used to push the air. The added electrical energy required to drive the fan may cancel out the extra solar heat gains.

ABSORBER COATINGS AND COVER PLATES

While considerations for absorber coatings, selective surfaces, and cover plates are similar for air and liquid collectors, there are a few differences. One of the primary drawbacks of a non-metalic absorber, such as in a plastic thin film collector, is the extreme difficulty of applying a selective surface to it. Until this technology improves, metal absorbers are preferred in applications where a selective surface is desirable. Low-cost, efficient air collectors will be readily available if selective surfaces can ever be applied to non-metal absorbers with ease.

As with liquid collectors, the use of a selective surface is about equivalent to the addition of a second cover plate. For Type I and II collectors, in which air flows between the absorber and the glazing, the addition of a second cover plate may be preferred because it creates a dead air space in front of the absorber.

The use of a "vee" corrugated absorber plate is somewhat analogous to the use of a selective surface. The vees create more surface area in the same square footage of collector area. It also increases the overall solar absorption (and hence the "effective" absorptance) because direct radiation striking the vees is reflected several times, with a little more absorption occurring

at each bounce. oriented properly, its absorptance is higher than that of a flat metal sheet coated with the same substance. But the increase in the emitted thermal radiation is small by comparison.

OTHER DESIGN FACTORS

Air leakage, though not as damaging as water leakage in a liquid collector, should be kept to a minimum. Because the solar heated air is under some pressure, it will escape through the tiniest crack. Prevention of air leakage helps to raise the collector efficiency. Take special care to prevent leakage through the glazing frames. By using large sheets instead of many small panes you can reduce the number of glazing joints and cut the possibility of leakage. And just as storm windows cut the air infiltration into your home, second and third cover plates reduce air leakage from a collector. Air leakage is the biggest factor in decaying efficiency and occurs throughout the system: collectors, ducts, and storage.

For Type I and II collectors, the turbulent flow through the air space in front of the absorber results in somewhat larger convection heat loss to the glass than is the norm with liquid collectors. Thermal radiation losses from the absorber are therefore a smaller part of the overall heat loss. The absorber in a Type I collector becomes relatively hot and loses a lot of heat out the back, so more insulation is required. But in Type II and III collectors, a turbulent air flow cools the back side somewhat and less insulation may be required.

One drawback of air as a heat transfer fluid is its low heat capacity. The specific heat of air is 0.24 and its density is about 0.075 pounds per cubic foot under normal conditions. By comparison, water has a specific heat of 1.0 and a density of 62.5 pounds per cubic foot. For the same temperature rise, a cubic foot of water can store almost 3500 times more heat than a cubic foot of air. It takes 260 pounds, or about 3500 cubic feet of air, to transport the same amount of heat as a cubic foot of water.

Because of this low heat capacity, large spaces through which the air can move are needed—even in the collector itself. Air passageways in collectors range from 1/2 to 6 inches thick. The larger the air space, the lower the pressure drop, but the poorer the heat transfer from absorber to air stream. And larger passages mean higher costs for materials. For flat, sheet-metal absorber surfaces, the passageway usually is 1/2 to 1 inch. Passageways ranging from 1-1/2 to 2-1/2 inches are standard for large collectors using natural convection or having unusually long (more than 15 feet) path lengths—the distance from the supply duct to the return duct.

13

Other Collector Types

Concentrating and *focusing collectors* may someday emerge as favorites. These collectors use one or more reflecting surfaces to concetrate sunlight onto a small absorber area. Collector performance is enhanced by the added sunlight hitting the absorber.

Depending upon their total area and orientation, flat reflectors can direct 50 to 100 percent more sunlight at the absorber. Focusing collectors only reflect direct sunlight onto the absorber. Concentrating collectors direct and diffuse radiation, so they also work well in cloudy or hazy weather—when diffuse sunlight is coming from the entire sky.

PARABOLIC COLLECTORS

Parabolic collectors have a reflecting surface *curved* to direct incoming sunlight onto a very small area. A deep parabolic surface (a fly ball hit to the outfield traces out a parabolic path) can focus sunlight on an area as small as a blackened pipe with fluid running through it. Such a focusing collector will perform extremely well in direct sunlight but will not work at all under cloudy or hazy skies because only a few of the rays coming from the entire bowl of the sky can be caught and reflected onto the blackened pipe. And even in sunny weather, the reflecting surface must pivot to follow the

sun so that the absorber remains at the focus. The mechanical devices needed to accomplish this tracking can be expensive and failure-prone. But the higher the temperatures and efficiencies possible with a focusing collector are sometimes worth this added cost and complexity for high-temperature applications.

COMPOUND PARABOLIC CONCENTRATOR

The compound parabolic concentrator was developed at the Argonne National Laboratory by physicist Dr. Roland Winston. His collector uses an array of parallel reflecting troughs to concentrate *both* direct and diffuse solar radiation onto a very small absorber—usually blackened copper tubes running along the base of each trough. The two sides of each trough are sections of parabolic cylinder—hence the name "compound parabolic concentrator" or CPC. Depending upon the sky condtion and collector orientation, a three- to eight-fold concentration of solar energy is possible. The collector performs at 50 percent efficiency while generating temperatures 150°F above that of the outside air.

The real beauty of the CPC collector is its ability to collect diffuse sunlight on cloudy or hazy days. Virtually all the rays entering a trough

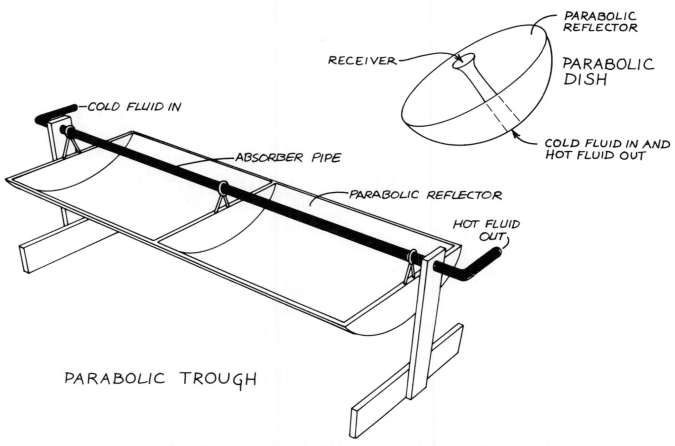

PARABOLIC TROUGH

Typical concentrating collector with parabolic reflectors. Direct rays from the sun are focused on the black pipe, absorbed and converted to heat.

are funneled to the absorber at the bottom. With the troughs oriented east-to-west, the collector need not track the sun. You merely adjust its tilt angle every month or so. After publishing his initial designs, Winston discovered that the same optical principles have been used by horseshoe crabs for thousands of centuries. These antediluvian creatures have a similar structure in their eyes to concentrate the dim light that strikes them as they "scuttle across the floors of silent seas."

EVACUATED-TUBE COLLECTORS

One of the biggest problems with flat-plate collectors is their large surface area for losing heat. Since the best insulator is a vacuum, a more-

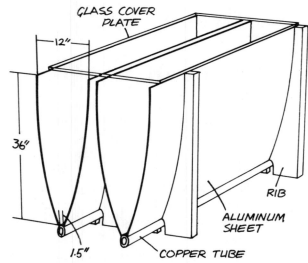

Compound parabolic collector invented by Dr. Roland Winston.

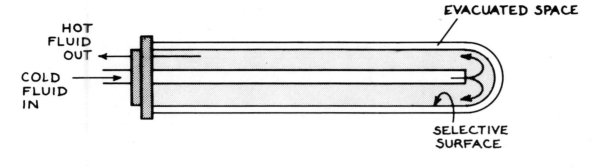

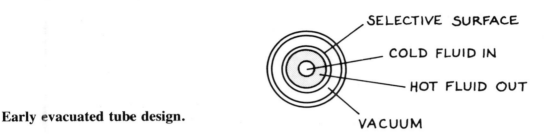

Early evacuated tube design.

efficient flat-plate collector would have a vacuum between its absorber plate and the cover sheet. The vacuum would eliminate the convective currents that steal heat from the absorber and pass it to the cover plate, which conducts it through to the outside. Better still would be an absorber with a vacuum on all six sides.

But vacuums cannot be created easily in a rectangular box without atmospheric pressure pushing in the cover. Researchers years ago applied the fluorescent tube manufacturing process to make solar collectors. One glass tube is placed within another and the space between them is evacuated—like a Thermos™ bottle. The inner tube has a selective coating on its outer surface, and an open-ended copper tube inside it, as shown in the figure. Air or water enters the copper tube from the header, and is forced out the open end of the copper tube, gathering heat absorbed in the inner glass tube before it returns to the header.

Most of the recent evacuated tube designs feature a U-shaped copper tube with a small selective-surface copper absorber. The copper plate absorbs the sun's energy and passes it to the heat transfer fluid flowing through the U-shaped tube.

Another design, called the heat-pipe evacuated tube, has a closed metal tube and plate inside the evacuated glass tube. The end of the metal tube, which extends just beyond the glass tube, protrudes into the header across the top of the collector. The refrigerant heat transfer fluid in the metal tube vaporizes when warmed by the sun, rises to the top of the tube, and condenses after conducting its heat to the water passing through the header. The condensed liquid falls down the side of the tube, to boil again.

The 3- to 4-inch diameter evacuated tubes are arranged side-by-side connected at one end to a header or heat exchanger (depending on the design) and supported at the other end. Evacuated-tubes are available for liquid, air, or phase-change systems. When pitched properly, they can even be used in drainback systems. No matter what the fluid or design, the collectors drastically cut heat loss from the absorber, and can have higher annual efficiencies than flat-plate collectors in cold, cloudy climates or in higher temperature applications.

Because the "cover plate" on an evacuated tube is a cylinder, less sunlight is reflected over the whole day than from a flat sheet of glass. And since many of the collector designs feature

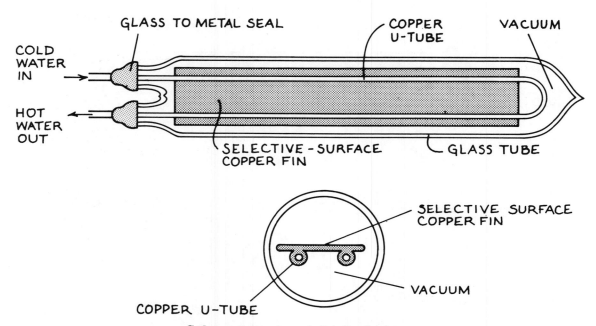

COLD WATER IN →

HOT WATER OUT ←

GLASS TO METAL SEAL

COPPER U-TUBE

VACUUM

SELECTIVE-SURFACE COPPER FIN

GLASS TUBE

SELECTIVE SURFACE COPPER FIN

VACUUM

COPPER U-TUBE

Subsequent evacuated tube design.

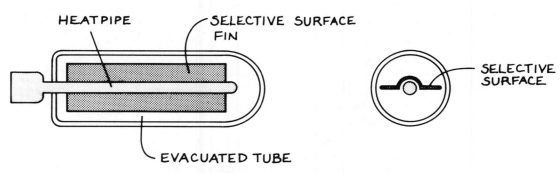

HEAT PIPE

SELECTIVE SURFACE FIN

SELECTIVE SURFACE

EVACUATED TUBE

Heat pipe evacuated tube.

flat or CPC-like reflectors underneath the tubes, they can collect diffuse as well as direct sunlight. This means they can collect energy on days when flat-plate collectors may be lying dormant.

Whether or not you need an evacuated-tube collector depends on your local climate and the temperature needed. In side-by-side tests, evacuated-tubes with CPC reflectors outperformed flat-plates during the winter, and performed about the same during the warm months, when their

lower heat losses aren't as important. This reveals the secret of concentrating collectors: it isn't their ability to concentrate energy that's important, but the fact that they have such small heat losses.

The two major drawbacks to evacuated-tube collectors are their high cost and tube breakage. Harder glass and better manufacturing processes and designs have reduced the second, but until larger production volumes are achieved, they will be more expensive than flat-plate collectors.

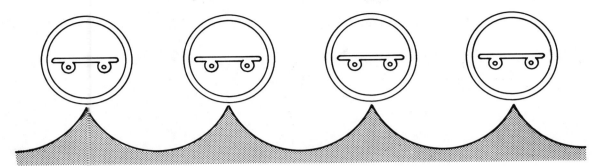

CPC evacuated-tube collector.

14
Collector Performance and Size

The performance of flat-plate collectors has been studied extensively. Most researchers try to predict the collector efficieny—the percentage of solar radiation hitting the collector that can be extracted as useful heat energy. A knowledge of the efficiency is very important in sizing a collector. If you know the available solar energy at your site, the average collector efficiency, *and* your heating needs, you're well on your way to determining the size of your collector.

The collector efficiency depends upon a number of variables—the temperature of the collector and outside air, the incoming temperature of the heat transfer fluid, the rates of insolation and fluid flow through the collector, and the collector construction and orientation. By manipulating the variables, a designer can improve overall collector performance. Unfortunately, few gains in efficiency are made without paying some penalty in extra cost. Beyond the obvious requirements of good collector location and orientation, many improvments in efficiency just aren't worth the added expense. Keep a wary eye turned toward the expenses involved in any schemes you devise to improve the efficiency.

COLLECTOR HEAT LOSSES

A portion of the sunlight striking the collector glazing never makes it to the absorber. Even when sunlight strikes a single sheet of glass at right angles, about 10 percent is reflected or absorbed. The maximum possible efficiency of a flat-plate glazed collector is therefore about 90 percent. Even more sunlight is reflected and absorbed when it strikes at sharper angles—and the collector efficiency is further reduced. Over a full day, less than 80 percent of the sunlight will actually reach the absorber and be converted to heat.

Further decreases in efficiency can be traced to heat escaping from the collector. The heat transfer from absorber to outside air is very complex—involving radiation, convection, and conduction heat flows. While we cannot hope to analyze all these processes independently, we *can* describe some important factors, including:

- average absorber temperature
- wind speed
- number of cover plates
- amount of insulation.

Perhaps you've already noticed that very similar factors determine the rate of heat escape from a house!

More heat escapes from collectors having hot absorbers than from those with relatively cool ones. Similarly, more heat escapes when the outdoor air is cold than when it is warm. The *difference*, in temperature between the absorber

114

Energy Flows in a Collector

Because energy never disappears, the total solar energy received by the absorber equals the sum of the heat energy escaping the collector and the useful heat energy extracted from it. If H_a represents the rate of solar heat gain (expressed in Btu/(ft² hr)) by the absorber, and H_e is the rate of heat escape, then the rate of useful heat collection (H_c) is given by:

$$H_c = H_a - H_e$$

Usually H_c and H_a are the easiest quantities to calculate, and H_e is expressed as the difference between them. The rate of solar heat collection is easily determined by measuring the fluid flow rate (R, in lb/(ft² hr)) and the inlet and outlet temperatures (T_{in} and T_{out}, in °F). The solar heat extracted, in Btu per square foot of collector per hour, is then:

$$H_c = (R)(C_p)(T_{out} - T_{in})$$

where C_p is the specific heat of the fluid—1.0 Btu/lb for water and 0.24 Btu/lb for air. Knowing H_c and the rate of insolation (I), you can immediately calculate the collector efficiency (E, in percent):

$$E = 100(H_c/I)$$

The instantaneous efficiency can be calculated by taking this ratio at any selected moment. Or an average efficiency may be determined by dividing the total heat collected over a certain time period (say an hour) by the total insolation during that period.

Of the total insolation the amount actually converted to heat in the absorber (H_a) is reduced by the transmittance (represented by the Greek letter tau, or τ) of the cover plates and by the absorptance (represented by the Greek letter alpha, or α) of the absorber. The value of H_a is further reduced (by 3 to 5 percent) by dirt on the cover plates and by shading from the glazing supports. Therefore, the rate of solar heat gain in the absorber is about

$$H_a = (0.96)(\alpha)(\tau)(I)$$

Both α and τ depend upon the angle at which the sunlight is striking the collector. Glass and plastic transmit more than 90 percent of the sunlight striking perpendicularly. But during a single day, the average transmittance can be as low as 80 percent for single glass, and lower for double glass. The absorptances of materials commonly used for collector coatings are usually better than 90 percent. If no radiation is converted to heat absorbed in the collector fluid, then

$$H_c = H_a = 0.96(\alpha)(\tau)(I)$$

and the average collector efficiency (for a whole day) would still be less than 80 percent. Unfortunately, there are large heat losses from a flat-plate collector, and efficiencies rarely get above 70 percent.

and the outdoor air, $\Delta T = T_{abs} - T_{out}$, is what drives the overall heat flow in that direction. The heat loss from a collector is roughly proportional to this difference.

As the absorber gets hotter, a point is eventually reached where the heat loss from a collector equals its solar heat gain. At this equilibrium temperature, the collector efficiency is zero—no useful heat is being collected. Fluids are usually circulated through a

collector to prevent this occurence. They carry away the accumulated heat and keep the absorber relatively cool. The higher the fluid flow rate, the lower the absorber temperature and the higher the collector efficiency.

Some fluids cool an absorber better than others. Although it has the disadvantages of freezing and corrosion, water is unmatched as a heat transfer fluid. It has the advantages of low viscosity and an extremely high heat capacity. Solutions of propylene glycol in water solve the freezing problems but they have a lower heat capacity. For the same flow rates, a 25 percent solution of glycol in water will result in a 5 percent drop in collector efficiency.

As a heat transport fluid, air rates a poor third. While optimum water flow rates are 4 to 10 pounds per hour for each square foot of collector, 15 to 40 pounds of air are usually needed. And the rate of heat transfer from absorber to fluid must also be considered. Rough corrugated surfaces work best in air collectors. Good thermal bonds and highly conductive metal absorbers are needed with liquid collector. The faster the heat transfer to the passing fluid, the cooler the absorber and the higher the collector efficiency.

As with houses, the collector heat losses can be lowered by adding insulation or extra glass. But extra glass also cuts down the sunlight reaching the absorber. The relation of these two factors to the collector efficiency is illustrated in the next two graphs. In the first, the equivalent of 2 inches of fiberglass insulation is placed behind the absorber. The back of the second absorber is very heavily insulated—so that virtually no heat escapes. In all cases, the daily average collector efficiency falls with increasing differences between the absorber and outside air temperatures. For absorber temperatures less than 40°F, the extra cover plates and insulation are obviously helpful. These remarks apply to a specific collector, but are generally true for most others. For very cold climates or very high fluid temperatures, the evacuated-tube collectors have the lowest heat losses.

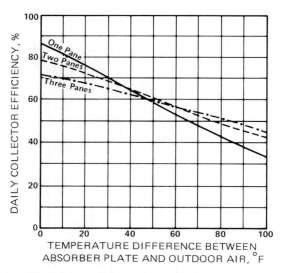

Performance of a moderately-insulated collector.

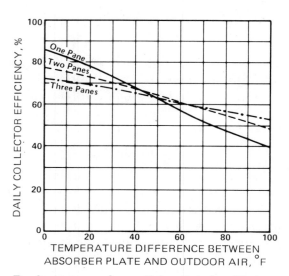

Performance of a well-insulated collector.

INSOLATION

The collector efficiency also depends upon the amount of sunlight hitting it. Under cloudy morning conditions, for example, the absorber will be much too cool to have fluid circulating through it, and no useful heat can be extracted. But at noon on a sunny day, the collector will be operating at full tilt, delivering 60 percent of the solar energy to storage. If the fluid flow can be increased to keep a constant absorber temperature, the collector efficiency will *increase* as the insolation rate increases.

The actual value of the insolation at a particular spot is very difficult to predict. Weather conditions vary by the hour, day, month, and year, and a collector designed for average condtions may perform quite differently at other times. For example, the collector described above may have an average daily efficiency of 40 percent, but its efficiency at any one moment can be anywhere in the range form 0 to 60 percent. Usually, we have to resign ourselves to using the *average* collector efficiency in our calculations. But that's not as bad as it may sound. The average daily insolation multiplied by the average collector efficiency gives us the solar heat collected per square foot on a typical day. With sufficient heat storage to tide us over times of shortage, why worry?

The Clear Day Insolation Data and the Percentage of Possible Sunshine Maps in the appendix provide a suitable method for calculating the insolation at most sites. They are limited to south-facing surfaces, unless you do a little trigonometry. They are capable of providing good estimates of the average insolation for any day and time. And fortunately a precise knowledge of the insolation isn't critical. Variations of 10 percent in the insolation will change collector efficiency by only 3–4 percent out of a total efficiency of 40 percent.

COLLECTOR ORIENTATION AND TILT

Two other factors that determine a collector's performance are its orientation and tilt angle. A collector facing directly into the sun will receive the most insolation. But flat-plate collectors are usually mounted in a fixed position and cannot pivot to follow the sun as it sweeps across the sky each day or moves north and south with the seasons. So the question naturally arises, "What is the best orientation and tilt angle for my collector?" In addition, designers need to know how much they can deviate from optimum.

Although true south is the most frequent choice for the collector orientation, slightly west of south may be a better choice. Because of early morning haze, which reduces the insolation, and higher outdoor air temperatures in the after-

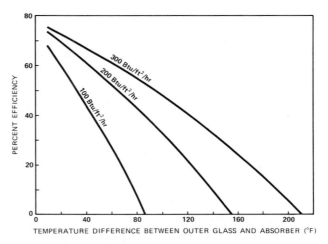

The effect of different insolation rates on collector efficiency. The outdoor air temperature is assumed constant.

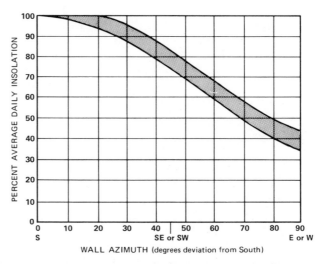

The percentage of insolation on vertical walls for orientations away from true south.

noon, such an orientation can give slightly higher collector efficiencies. On the other hand, afternoon cloudiness in some localities may dictate an orientaion slightly east of south. Fortunately, deviations of up to 15°F from true south cause relatively small reductions in collector efficiences. The designer has a fair amount of flexibility in his choice of collector orientation.

A useful diagram shows the approximate decrease in the insolation on a vertical wall col-

BOSTON, MASSACHUSETTS

CHARLESTON, SOUTH CAROLINA

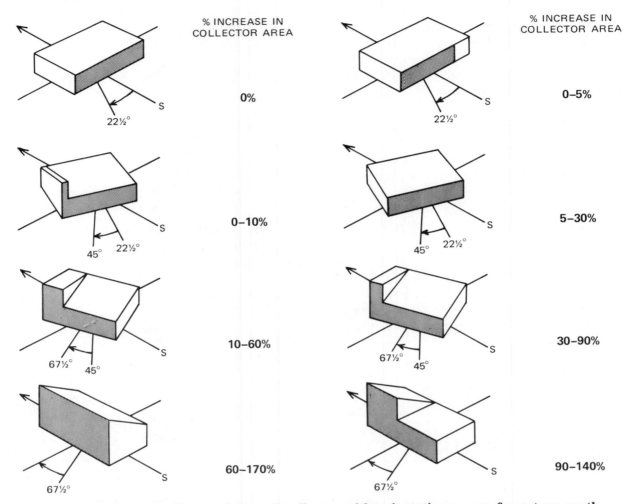

The change in area of a vertical wall collector with orientations away from true south. The collector (shaded areas) has been sized to provide 50 percent of the winter heating needs of a well-insulated home in Boston and Charleston.

lector facing away from true south. The graph is valid for latitudes between 30°N and 45°N—almost the whole United States, if we forget Alaska and Hawaii. And it applies to the coldest part of the year—from November 21 to January 21. You can use this graph together with the Clear Day Insolation Data to get a rough estimate of the clear day insolation on surfaces that do not face true south. Simply multiply the data by the percentage appropriate to the orientation you have selected.

The effect orientation has upon the required size of a collector is illustrated by two examples in the next diagram. The vertical wall collectors in all cases are sized to provide 50 percent of the heating needs of a 1000-square-foot house in either Boston or Charleston, South Carolina. Note that southwest (or, for that matter, southeast), orientations require an extra collector area of only 10 percent in Boston and 30 percent in Charleston. Tilted surfaces are affected even less by such variations.

118

The collector tilt angle depends upon its intended use. A steeper tilt is needed for winter heating than for summer cooling. If a collector will be used the year round, the angle chosen will be a compromise. If heating and cooling needs do not have equal weight, your selection of a tilt angle should be biased toward the more important need.

The general consensus holds that a tilt angle of 15° greater than the local latitude is the optimum for winter heating. For year-round uses, the collector tilt angle should equal the local latitude. And for maximum summer collection, the best choice is 15° less than the latitude. But your house won't freeze up or boil over if you don't have exactly the right tilt angle. The collector can be tilted 15–20° away from the optimum and still get more than 90 percent of the maximum possible insolation.

For areas with severe winter cloudiness, steeper tilts may be required. Sometimes vertical wall collectors are a good idea. Such a collector receives its peak insolation in the winter months, but gets very little in the summer, so don't expect to heat a swimming pool with it. When reflections from the ground are added to the sunlight hitting a vertical collector directly, its performance can surpass that of a collector tilted at the "optimum" winter heating angle (latitude + 15°). Clean, fresh snow has the highest reflectance—87 percent—of any common surface. It can add another 15 to 30 percent to the solar heat output of a vertical collector. Other surfaces such as asphalt, gravel, concrete, and grass have reflectances ranging from 10 to 33 percent.

The accompanying diagram shows the relationship between tilt angle and collector size. The collectors in all four cases are sized to provide 50 percent of the January heating load for a typical Minneapolis home oriented true south. Relatively small changes in collector size can compensate for the reductions in solar collection resulting from changes to the January "optimum" (50°) tilt angle. With additional sunlight reflected from fallen snow, a vertical wall collector can be 14 percent *smaller* than the rooftop collector tilted at 50°.

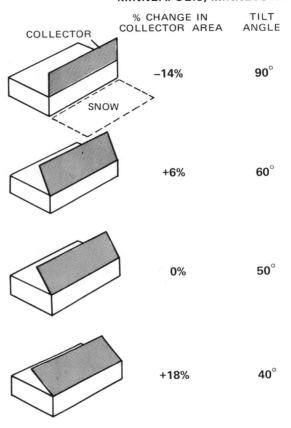

MINNEAPOLIS, MINNESOTA

COLLECTOR	% CHANGE IN COLLECTOR AREA	TILT ANGLE
	−14%	90°
	+6%	60°
	0%	50°
	+18%	40°

NOTES: 1. Base tilt angle = 50°.
2. Operating temperature = 90°F.
3. Due South orientation.

Small changes in collector size are required when the tilt angle differs from the optimum.

SIZING THE COLLECTOR

Accurate performance predictions are needed for sizing a collector. It is important to know whether 35 or 40 percent collector efficiencies (on the average) can be expected in any given month. With this knowledge and some predictions of the average daily insolation for that month, you'll have a good idea of how much solar heat you can expect from each square foot of collector. The size then follows from the average monthly heat demands.

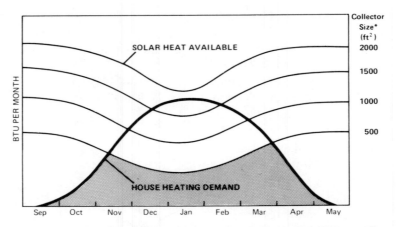

The solar heat delivered per square foot of collector decreases as the collector size increases. The shaded area
represents the portion of the heating demand supplied by
a 500-square-foot collector. Doubling the collector size may
double the solar heat supplied in January, but it doesn't
help much in October or April.

But the performance of a collector is even
more difficult to predict when it is tied into an
entire heating system. Although some rules of
thumb have emerged, the subject is still clouded
in mystery as far as the average homeowner or
builder is concerned. Site conditions, the heating demands of the house, and specific design
choices (such as the collector tilt, the operating
temperature, and the heat storage capacity) all
affect the average collector performance.

A well-designed collector might be able to
collect 1200 Btu per square foot on a sunny
winter day. But not all of this heat will reach
the rooms. There will be heat losses from the
heat storage container, which may already be
too hot to accept additional heat. There will also
be heat lost from the ducts or pipes between the
collector and the storage tank and the rooms.
Solar energy will be rejected during extended
periods of sunny weather—even if the outdoor
temperatures are quite cold. Only when the proper
sequence of sunny and cloudy days occurs can
all the available solar energy be used.

Consider a 1000-square-foot house with a
heating demand of 12,000 Btu per degree day,
or 84-million Btu in a 7000-degree-day climate.
This demand is distributed over the heating sea-

son (late September to early May) as shown in
the diagram. Little heat is needed in October or
April and the bulk of the heat is needed from
December through February—just when sunlight is at a premium.

Such a house needs about 600,000 Btu on a
day when the outside temperature averages 15°F.
On a sunny day, this amount of solar heat can
be supplied by 500 square feet of a good collector. And with a 35°F temperature rise (from
85°F to 120°F, for example), about 2000 gallons
of water will absorb all of the solar heat. But
under *average* operating conditions, a square
foot of this collector gains only 350 Btu of usable solar heat per day—or 84,000 Btu for the
entire seven-month heating season. The 500
square feet of collector will supply only half of
the seasonal heating load of 84 million Btu.

Contrary to what you might expect, doubling
the collector size from 500 to 1000 square feet
does *not* provide 100 percent of the heating
load. Instead, it provides about 75 percent because the larger collector does not work to full
capacity as often as the smaller one. In this
particular system, the usable heat per square
foot of collector drops from 84,000 Btu to 69,000
Btu because the house just cannot use the added

COLLECTOR SIZE AND SOLAR HEAT DELIVERED

Collector Size* (ft^2)	Solar Heat Used per Square Foot of Collector	Total Solar Heat Supplied per Season MMBtu$^+$	of Demand
2000	42,000	84	100%
1500	56,000	75	89%
1000	59,000	63	75%
500	84,000	42	50%

*Storage size remains fixed. $^+$1 MMBtu = 1 million Btu.

heat in the fall and spring. And the system must reject more heat during four or five successive days of January sunshine. Even if the storage size were doubled, the system would have to reject excess heat during the autumn and spring. As collector size increases for a fixed heat demand, the amount of solar energy provided by each square foot drops because of the decreased *load factor* on each additional square foot. This is the law of *diminishing returns*.

A simplified method for collector sizing, lies somewhere between educated guess-work and detailed analytical calculations. It is accurate to within 20 percent and is biased toward conservative results. Architects, designers, and owner-builders often need such a "first-cut" estimate of collector size to proceed with their designs.

First the monthly output of a flat-plate collector is calculated as the product of the *useful* sunshine hours in the month times the average hourly solar heat output of the collector during the month. The number of useful sunshine hours is less than the total number of sunshine hours because the insolation rate is not high enough in early morning to justify collector operation. The two most influential factors that determine the hourly solar output—the hourly insolation rate and the temperature difference between the absorber and the outside air—change rapidly during a single day. Therefore, we only try to determine an *average* hourly solar output. The method outlined in "Estimating Collector Performance" is an attempt to determine the reasonable mean monthly output from average insolation rates and temperature differences. More detailed analyses are usually made using the F-chart computer simulations described earlier.

A sample calculation of mean monthly solar heat output is provided for illustration. The hypothetical collector is sited in Boston and oriented south at a tilt angle of 60°. At an average operating temperature of 120°F, this collector has an efficiency of 38 percent and gathers 9880 Btu per square foot during the month of January.

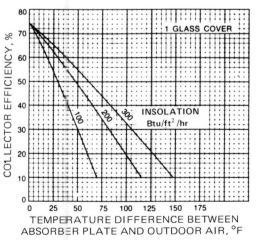
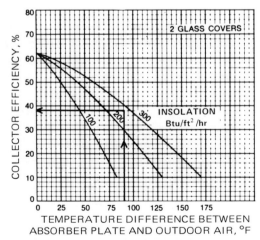

Performance curves for single- and double-glazed collector. Use these graphs to estimate collector efficiency from the insolation rate and temperatures of the absorber and outdoor air. SOURCE: Revere Copper and Brass Co.

Estimating Collector Performance

The following method will help you estimate the average efficiency and monthly solar heat output of a well-built solar collector. This method is accurate to about 20 percent and results in conservative estimates of performance. A running example is included for illustration.

1. *Find the total number of hours of sunshine for the month from the "Mean Number of Hours of Sunshine" table in the appendix; for example, 148 hours for Boston in January.*
2. *Find the average day length for the month from almanacs or the weather bureau; 10 hours for Boston in January.*
3. *From the Clear Day Insolation tables in the appendix, calculate the number of "collection hours" per day. For the selected collector tilt angle, this is the number of hours that the insolation is greater than 150 Btu/ft^2. Count 1/2 hour for insolation rates between 100 and 150 Btu/ft^2; 7 hours for 40°N latitude and 60° tilt.*
4. *Determine the total collection hours per month by multiplying the sunshine hours per month (#1) by the collection hours per day (#3) and dividing by the average day length (#2); 148 hours (7 hours) ÷ 10 hours = 104 hours.*
5. *Determine the total daily useful insolation, defined as the total insolation during collection hours, by adding the hourly insolation rates (from the Clear Day Insolation tables) for those collection hours described above; 187 + 254 + 293 + 306 + 293 + 254 + 187 = 1774 Btu/(ft^2 day).*
6. *Determine the average hourly insolation rate (I) during the collection period by dividing the total daily useful insolation (#5) by the number of collection hours per day*

(#3); 1774 Btu/(ft^2 day) ÷ 7 hours/day = 253 Btu/(ft^{2}hr).

7. *Determine the average outdoor temperature (T$_a$) during the collection period from 1/2 the sum of the normal daily maximum temperature and the normal daily average temperature for the month and locale. These are available from the local weather bureau and from the Climatic Atlas of the United States . T$_{out}$ = 1/2(38°F) + (30°F) = 34°F.*
8. *Select the average operating temperature (T$_{abs}$) of the collector absorber and find the difference (ΔT = T$_{abs}$ − T$_{out}$). In general, you should examine a range of possible values for T$_{abs}$: ΔT = 120°F − 34°F = 86°F.*
9. *Refer to a performance curve for the collector to determine the average collector efficiency from a knowledge of ΔT (#8) and I (#6). The sample curves provided apply to a tube-in-plate liquid-type collector, but they should be fairly accurate for most flat-plate collectors of moderate to good construction; average collector efficiency = 38 percent for a double-glazed collector.*
10. *Determine the average hourly collector output by multiplying the average hourly insolation rate (I from #6) by the average collector efficiency (#9); 0.38(253 Btu/(ft^2 hr)) = 96 Btu/(ft^2 hr).*
11. *The useful solar heat collected during the month is then the average hourly collector output (#10) multiplied by the number of collection hours (#4) for that month: 96 Btu/(ft^{2}hr)(104 hours/month) = 9984 Btu/ft^2 for January in Boston.*

This procedure should be repeated for a number of other collector operating temperatures and tilt angles.

ESTIMATES OF MONTHLY COLLECTOR OUTPUT (in Boston)

Collector		Average Solar Heat Collected (Btu/ft^2)									
°F	Tilt	Sep	Oct	Nov	Dec	Jan	Feb	Mar	Apr	May	TOTALS
90	60°	21,700	19,630	13,780	13,080	12,480	14,640	18,000	14,720	15,225	143,255
90	90°*	14,615	19,781	14,310	14,170	13,000	15,250	10,925	5,040	2,520	109,611
120	60°	18,600	15,855	11,130	9,810	9,984	11,590	14,250	12,160	12,325	115,704
120	90°*	12,700	16,006	11,660	11,455	10,400	12,200	8,625	3,240	1,800	88,076
140	60°	16,275	12,835	9,010	8,175	7,800	9,150	11,250	10,240	9,425	94,160
140	90°*	10,160	12,986	9,540	9,265	8,320	9,750	5,750	2,160	1,080	69,021

*With 20 percent ground reflection.

Monthly solar output for the rest of the heating season has been calculated with the same method and listed in the accompanying table. The output of a vertical collector (including 20 percent ground reflectance) is included in the table, as are the monthly outputs when 90°F and 140°F operating temperatures are allowed. The seasonal output is the sum of all these monthly figures. In your design work, it's extremely useful to consider a number of alternative collector tilts and operating temperatures—instead of proceeding single-mindedly with a preconceived design. Almost every collector operates over a range of temperatures and its efficiency varies in a corresponding fashion. It's instructive to determine the solar heat collection for a few of these conditions.

In general, the larger the percentage of house heating you want your collector to supply, the more difficult it is to estimate its size using these simplified methods. The actual sequence of sunny and cloudy days becomes more important as the percentage of solar heating increases. If a full week of cold, cloudy days happens to occur in January, your collector (or storage) would have to be enormous to insure 90 percent solar heating. But good approximations of collector size can be made for systems that are designed to supply 60 percent or less of the seasonal heating needs.

A simplified method of calculating the collector size from monthly output and heating demand figures is outlined in "Estimating Collector Size." The monthly output figures are those of our hypothetical collector, tilted at 60° and operating at an absorber temperature of 120°F. The heating demand figures are for a Boston home of 1000 square feet that loses 9500 Btu per degree day. In this particular example, we strive to provide 50 percent of the seasonal heat demands of the house. If the initial guess at the appropriate collector size does not provide the desired percentage, it can be revised up or down and the calculations repeated until the desired results are achieved.

The final size of the collector should reflect other factors besides heating demand—for example, the size of the heat storage container, the solar heat gain through the windows, available roof or wall area, and cost.

COMPARING COLLECTORS

Some states require that manufactured solar collectors be tested and rated on how much thermal energy they produce. The Solar Rating and Certification Corporation (SRCC) is a non-profit organization incorporated in 1980 to develop

The New Solar Home Book

Estimating Collector Size

The following procedure helps you to estimate the collector size needed to supply a desired percentage of the yearly heating demand. To use it, you need the monthly output per square foot of collector, as calculated in "Estimating Collector Performance." The Boston example is continued here for illustration—with the collector tilted at 60° and operating at 120°F.

1. *For the tilt angle and operating temperature selected, enter the monthly output per square foot of collector in column A. Add them to get the heating season output for one square foot.*
2. *Enter the monthly degree days of the location (from the "Degree Days and Design Temperatures" table in the appendix) in column B.*
3. *Enter the monthly heat loss of the house in column C. This is the product of the monthly degree days times the heat loss per degree day—or 9500 Btu per degree day for our Boston home.*
4. *Add the entries in column C to determine the seasonal heat loss. Divide this total by the total of column A (step 1) and take 60 percent of the result as a first guess at the collector area needed to supply 50 percent of the seasonal heat demand: 0.60(53.46*

MMBtu)(100,000Btu) ÷ 115,704 Btu/ft² = 277.2 ft².

5. *Multiply this collector area by the entries in column A and enter the resulting solar heat collected in column D.*
6. *Subtract entries in column D from those in column C to obtain the heat demand NOT met by solar energy during the month. If a negative result occurs, solar energy is supplying more than can be used, and a zero should be recorded in column E.*
7. *Subtract entries in column E from those in column C to get the total solar heat used by the house in the month. Enter these results in column F.*
8. *Divide entries in column F by those in column C and multiply by 100 to get the percentage of monthly heat losses provided by solar (column G).*
9. *Divide the seasonal total of column F by that of column C to get the percentage of the seasonal heat loss provided by solar, or 47 percent in the Boston home. If this result is too low (or high) the collector area can be revised in step 4 and steps 5 to 9 repeated until satisfaction is achieved.*

The total of column F is the "useful" solar energy output of the collector. It can be used to predict the economic return on the initial expenses of the system.

Month	A Collector Output (Btu/ft²)	B Degree Days (°F days)	C Heat Loss (MMBtu*)	D Solar Heat Collected (MMBtu*)	E Auxiliary Heat (MMBtu*)	F Solar Heat Used (MMBtu*)	G Percent Solar Heated
September	18,600	98	0.93	5.16	0	0.93	100
October	15,855	316	3.00	4.40	0	3.00	100
November	11,130	603	5.73	3.09	2.64	3.09	54
December	9,810	983	9.34	2.72	6.62	2.72	29
January	9,884	1,088	10.34	2.74	7.60	2.74	26
February	11,590	972	9.23	3.21	6.01	3.21	35
March	14,250	846	8.04	3.95	4.09	3.95	51
April	12,160	513	4.87	3.37	1.50	3.37	69
May	12,325	208	1.98	3.42	0	1.98	100
Heating Season Totals	115,704	5,627	53.46	32.06	28.46	24.99	47

*Millions of Btu. House Heat Loss: 9500 Btu/deg day. Collector Area: 277.2ft².

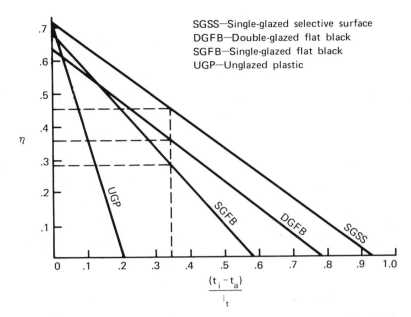

SGSS—Single-glazed selective surface
DGFB—Double-glazed flat black
SGFB—Single-glazed flat black
UGP—Unglazed plastic

Four sample thermal efficiency curves. (*Solar Age*)

and implement certification programs and national rating standards for solar equipment. The SRCC's collector certification program provides a means to evaluate the maintainability, structural integrity, and thermal performance of solar collectors under strict laboratory conditions. The tests, paid for by each manufacturer, are conducted by independent laboratories accredited by the SRCC.

Collectors or whole systems are randomly selected and inspected upon receipt to check the original condition after shipping. The collector then undergoes a pressure test to see if it leaks, and is exposed to the weather for 30 days. After exposure, the collector is checked for signs of degradation. A series of tests, from thermal shock to thermal performance, is conducted before the collector is taken apart and inspected one last time.

The thermal performance test determines the instantaneous efficiency of the collector. With the outside air temperature and incident solar radiation level held constant, the inlet temperature is varied four times to see how well the collector operates in four different temperature ranges. The data collected is plotted on the collector's *thermal efficiency curve*. The curve helps you compare the instantaneous efficiencies of

different collectors, so that with the cost of each collector, you can decide which collector is right for your location and application.

The figure shows the thermal efficiency curves for four collectors: an unglazed collector with a plastic absorber, a single-glazed collector with a flat-black painted absorber, a double-glazed collector with a flat-black absorber, and a single-glazed collector with a selective-surface absorber.

Following our last example, if the average insolation rate (I) in January is 253 Btu/(ft^{2}hr) in Boston, the average daytime temperature (T$_a$) is 34°F, and the collector inlet temperature (T$_i$) is 120°F, you can use the thermal efficiency curves to find each collector's average instantaneous efficiency in this application. The fluid parameter, plotted along the x-axis, is equal to the inlet temperature (T$_i$) minus the daytime temperature (T$_a$), all divided by the insolation rate (I) :

Fluid parameter = (T$_i$ − T$_a$) / I

In this case, the fluid parameter equals (120 − 34) /253, or 0.34. Starting at that point on the x-axis, mark the intersections with the efficiency curves, and read the efficiency from the y-axis. The efficiency for the single-glazed selective

SGSS—Single-glazed selective surface
DGFB—Double-glazed flat black
SGFB—Single-glazed flat black
UGP—Unglazed plastic

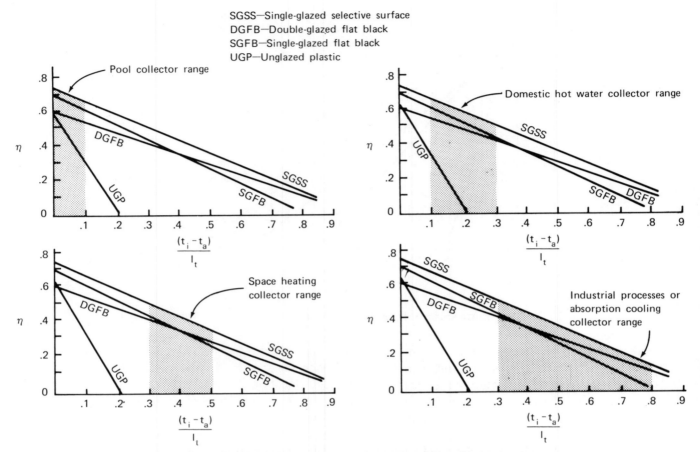

General fluid parameter boundaries for different applications. (*Solar Age*)

surface collector if 0.45, for the double-glazed flat-black collector is 0.36, and for the single-glazed flat-black collector is 0.28. The unglazed plastic collector cannot compete in this range —it only performs well in low-temperature applications, such as pool heating or domestic hot water in very warm climates.

The thermal efficiency curve can tell you a lot about the collector. The point where it intersects the left side of the graph represents the maximum efficiency the collector can achieve (at that point its losses equal zero). If the manufacturer tells you his collector can deliver half the energy it receives, and its collector efficiency curve intersects the y-axis at 0.50, he's exaggerating. The curve doesn't account for the rest of the system losses!

The steeper the slope of the curve, the less efficient the collector is at higher temperatures. As the fluid parameter increases, collector efficiency decreases. Swimming pool collectors have the steepest slopes because their losses are high. They are best suited in the fluid parameter range below 0.10 (see figure). Collectors for solar domestic hot water have less steep slopes so that they can collect energy better in the 0.10 to 0.30 fluid parameter range. Space heating collectors are made for the 0.30 to 0.50 range, and industrial process heating or absorption cooling collectors need to perform well in the 0.3 to 0.8 fluid parameter range.

Single-glazed flat-black collectors have steeper slopes than double-glazed collectors since their losses are higher. But selective surfaces on single-glazed, double-glazed, or evacuated-tube collectors outperform the rest since their radiation losses are cut significantly.

The instantaneous efficiencies only help to compare collectors and shouldn't be used to determine annual performance, since it is only the *instantaneous* efficiency under optimum conditions. It doesn't account for the difference in collector efficiency at the beginning or end of the day versus that at noon, when the sun's rays are more perpendicular to the absorber and insolation rates are higher. It doesn't say what happens under hazy skies. In both cases, collectors such as evacuated-tubes can perform better than flat-plates. Another drawback to the collector test is that it only tests the collector efficiency, and doesn't subtract the tank or distribution losses or pumping power required.

The SRCC tests for complete systems do take into account losses from the tank and how much pump power is used. Tank losses at night are important if you're comparing an active system (with the tank in a "heated" basement) to an integral storage system (with its tank exposed to the cold night air). Taking pump power into account is important if you're comparing passive and active systems. But remember again that the results you see are only those gathered under strict laboratory conditions, and not what you'll get for the same systems in your location and your operating conditions.

Comparing efficiency curves is like comparing EPA automobile gas milage ratings. There are many other things that should be looked at before you decide which collector to buy: cost, appearance, and anticipated lifetime.

15
Storage and Distribution

Some capacity for storing solar heat is almost always necessary because the need for heat continues when the sun doesn't shine. And more heat than a house can use is generally available when the sun *is* shining. By storing this excess, an active system can provide energy as needed—not according to the whims of the weather.

If costs were not a factor, you would probably design a heat storage unit large enough to carry a house through the longest periods of sunless weather. A huge, well-insulated storage tank (say 15-20,000 gallons) in the basement could store heat from the *summer* for use in *winter*! But most of us do not have the money to spend on an enormous storage tank, and our designs are limited by what we can afford. Some of the major factors influencing heat storage costs are:

- choice of storage medium
- amount or size of the storage medium
- type and size of a container
- location of the heat storage
- use of heat exchangers, if required
- choice of pumps or fans to move the heat transfer fluid

In designing solar heat storage, you must weigh these cost considerations against the performance of the system. All of the above factors influence performance to some extent. Other factors include the average operating temperature of the entire system, the pressure drop of the heat transfer fluid as it passes through or by the storage medium, and the overall heat loss from the container to its surroundings.

In general, the heat storage capacity of common storage materials varies according to their specific heat—the number of Btu required to raise the temperature of 1 pound of a material 1°F. The specific heats of a few common heat storage materials are listed in the table along with their densities and heat capacities—the amount of heat you can store in a cubic foot of the material for a 1°F temperature rise. Heat energy stored with an accompanying rise in temperature is called *sensible heat*, and it is reclaimed as the temperature of the storage medium falls. It takes high temperatures or large volumes of material to store enough sensible heat (say 500,000 Btu) for a few cold, sunless days. Rocks and water are by far the most common storage media because they are inexpensive and plentiful.

Some materials absorb a lot of heat as they melt, and surrender it as they solidify. A pound of Glaubers salt absorbs 104 Btu and a pound of paraffin 65 Btu when they melt at temperatures not far above normal room temperatures. The heat absorbed by the change in phase from liquid to solid and solid to liquid is called latent

PROPERTIES OF HEAT STORAGE MATERIALS

Material	Specific Heat [Btu/(lb °F)]	Density [lb/ft^3]	Heat Capacity [Btu/(ft^3 °F)]	
			No voids	30% voids
Water	1.00	62	62	43
Scrap Iron	0.12	490	59	41
Scrap Aluminum	0.21	171	36	26
Concrete	0.22	144	32	22
Stone	0.21	170	36	25
Brick	0.20	140	28	20

heat, and is stored or released without a change in temperature. For example, when one pound of ice melts, it absorbs 144 Btu, but stays at 32°F. This is latent heat storage. To raise its temperature to 33°F, it takes 1 Btu—sensible heat storage.

The storage of heat in phase-changing materials can reduce heat storage volumes drastically. But continuing problems of cost, containment, and imperfect re-solidification have limited their use.

A solar collector and heat storage medium should be chosen together. Liquid collectors almost always require a liquid storage medium. Most air collectors require a storage medium consisting of small rocks, or small containers of water or phase-change materials. These allow the solar heated air to travel around and between—transferring its heat to the medium. Within these basic categories of heat storage, there are many possible variations.

TANKS OF WATER

Water is cheap and has a high heat capacity. Relatively small containers of water will store large amounts of heat at low temperatures. From 1 to 2 gallons are needed per square foot of solar domestic hot water collectors, and 1 to 10 gallons are needed per square foot of space heating collectors—or 500 to 5000 gallons for a 500-square-foot space heating collector. An-

other advantage of water heat storage is its compatibility with solar cooling. But there are several problems with water storage, such as the high cost of tanks and the threat of corrosion and possible leakage.

Water containment has been simplified in recent years by the emergence of good waterproofing products and large plastic sheets. Previously, the only available containers were leak-prone galvanized steel tanks. Their basement or underground locations made replacement very difficult and expensive. Glass linings and fiberglass tanks helped alleviate corrosion problems but increased initial costs. Until recently, the use of poured concrete tanks has been hampered by the difficulty of keeping them water tight—concrete is permeable and develops cracks. But large plastic sheets or bags now make impermeable liners having long lifetimes. And with lightweight wood or metal frames supporting the plastic, the need for concrete can be eliminated.

The most straightforward heat storage system (see diagram) is a water-filled container in direct contact with both the collector and the house heating system. The container shown is made of concrete or cinder blocks with a waterproof liner, but it might well be a galvanized or glass-lined tank. The coolest water from the bottom of the tank is circulated to the collector for solar heating and then returned to the top of the tank. Depending upon the time of day, the temperature difference between the bottom and top of

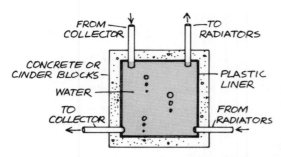

Heat storage tank is tied directly to both the collector and the house heating system in an open-loop system.

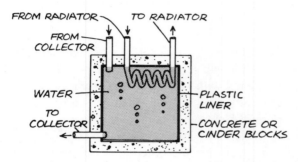

Use of a heat exchanger to extract solar heat from a storage tank, in a closed-loop system.

a 3-to 4-foot high tank can be 15 to 24°F. Sending the coolest water to the collectors improves collector efficiency. In an open-loop system, the warmest water from the top of the tank is circulated directly through baseboard radiators, a water-to-air heat exchanger, or radiant heating panels inside the rooms.

If the system is a closed loop, it might have a heat exchanger—a copper coil or finned tube—immersed in the tank of solar-heated water. Water or another liquid circulates through the heat exchanger, picks up heat, and carries it to the house. Warm water in the tank can also be pumped through heat transfer coils located in an air duct. Cool room air is blown past the coils and heated.

Heat exchangers are necessary when the water in the tank cannot be used for purposes other than heat storage. For example, an antifreeze solution used in a solar collector is often routed through a heat exchanger to prevent mixing with

water in the tank. And heating engineers often insist that water in the tank *not* be used in the room radiators—particularly when the tank water is circulated through the collector—because of corrosion. Because of their large size, some of these heat exchangers can be expensive. For a typical metal heat exchanger submerged in the water tank, the total metal surface can be as much as 1/3 the surface area of the solar collector.

For the designer who wishes to include heat storage as an integral part of a total design, the placement of a large unwieldly tank can be a problem. Self-draining systems require a tank located below the bottom of the collector, and thermosiphoning systems need it above the collector top. If the storage tank is linked to other equipment such as a furnace, pumps, or the domestic water heater, it will probably have to be located near them.

One-gallon or smaller containers of water can and have been used as the heat storage medium in air systems. They are arranged in racks, on shelves, or in any fashion that allows an unobstructed air flow around them. Possible containers include plastic, glass or aluminum jars, bottles, or cans.

ROCK BEDS

Rocks are the best known and most widely used heat storage medium for air systems. Depending upon the dimensions of the storage bin, rock diameters of 1 to 4 inches will be required. But through much of New England, for example, the only available rock is 1- to 1 1/2 -inch gravel. Even if the proper size is available, a supplier may be unable or unwilling to deliver it. Collecting rocks by hand sounds romantic to the uninitiated but becomes drudgery after the first thousand pounds. And many thousands of pounds—from 100 to 400 pounds per square foot of collector—are required because of the low specific heat of rock.

A large storage bin must be built to contain the huge quantities of rock needed. With 30 percent void space between the rocks, the bin requires about 2 1/2 times as much volume as a tank of water to store the same amount of heat over the same temperature rise. And the large surface area of rock storage bin leads to greater heat losses.

Rock storage bins can be used in systems which combine cooling and domestic water heating with space heating. Cool night air is blown over the rocks, and the coolness stored for daytime use. To preheat domestic hot water, cold water from city mains can pass through a heat exchanger located in the air duct returning from the collector to the storage bin.

The location of a rock storage bin must take into account its great volume and weight. It can be located in a crawl space under the house or under a poured-concrete slab at a small additional cost. Putting it inside the basement or other living space is usually more difficult and expensive.

To distribute heat from a rock storage system, air is either blown past the hot rocks or allowed to circulate through them by gravity convection. From there, the air carries solar heat to the rooms. In general, a fan or a blower is needed to augment the natural circulation and give the inhabitants better control of the indoor temperature.

A basic method of transferring heat to and from a heat storage bed is shown schematically in the diagram. Solar heated air from the collector is delivered to the *top* of the bin. It is drawn down through the rocks and returns to

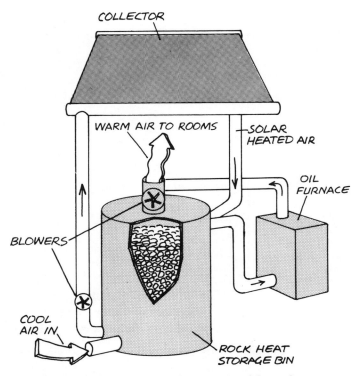

Schematic diagram of an air system with rock heat storage.

the collector from the bottom of the bin. To heat the house, cool air is drawn in at the bottom and is heated as it rises through the warm rocks. The warmest rocks at the top transfer their heat to the air just before it is sent to the rooms. The furnace heating cycle (also shown) draws even slightly pre-heated air from the top of the rock storage, boosts it to the necessary temperature, and delivers it through the same ductwork to the house. The furnace is placed in line *after* the rock bed.

Solar heated air is brought in at the top of a rock storage bed in order to encourage temperature stratification. House air can then be heated to the highest possible temperature by the warmest rocks at the top. But if solar heated air comes in at the bottom, the heat percolates upward and distributes itself evenly through the entire bed —resulting in lower temperatures throughout. Bringing cool room air in at the warm top also promotes this unwanted even heat distribution.

The shape of a rock storage bin is closely related to rock size. The farther the air must travel through the rocks, the larger the rock

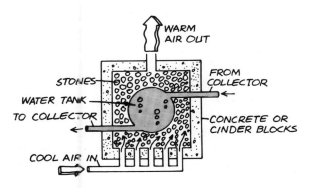

This approach uses both water and stone as storage media.

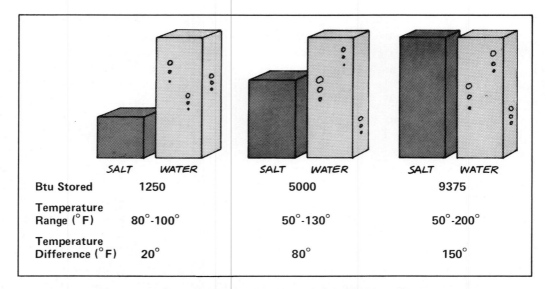

	SALT	WATER	SALT	WATER	SALT	WATER
Btu Stored	1250		5000		9375	
Temperature Range (°F)	80°-100°		50°-130°		50°-200°	
Temperature Difference (°F)	20°		80°		150°	

The volume of Glaubers salt needed to store the same amount of heat as a cubic foot of water. The salt volume indicated includes 50 percent voids between the containers of salt.

diameter required to keep the pressure drop and fan size small. If the path length through the rock bed is more than 8 feet, the rocks should be at least 2 inches in diameter—and larger for longer paths. For shorter path-lengths 1- to 2-inch gravel can be used.

The optimum rock diameter depends a lot on the velocity of the air moving through the rocks. The slower the air speed, the smaller the rock diameter or the deeper the bed of rocks can be. And the smaller the rock diameter, the greater the rock surface area exposed to the passing hot air. A cubic foot of 1-inch rock has about 40 square feet of surface area while the same volume of 3-inch rock has about 1/3 as much. In general, the rocks, stones, gravel, or pebbles should be large enough to maintain a low pressure drop but small enough to insure good heat tranfer.

PHASE-CHANGE MATERIALS

Phase-change materials, such as eutectic salts, are the only real alternative to rocks and containers of water as the heat storage for an air system. A eutectic salt absorbs a large amount of heat as it melts at a low temperature and releases that heat as it solidifies. A pound of Glaubers salt, the most widely studied and used, absorbs 104 Btu as it melts at 90°F and about 21 Btu as its temperature rises another 30°F. To store the same 125 Btu in the same temperature range requires about 4 pounds of water or 20 pounds of rocks.

Much smaller storage volumes are possible with eutectic salts. Consequently, they offer unusual versatility in storage location. Closets, thin partitions, structural voids, and other small spaces within a house become potential heat storage bins.

But this advantage is less pronounced when you increase the temperature range over which the salt cycles. The diagram illustrates the volume of Glaubers salt needed to store the same amount of heat as a cubic foot of water over three different temperature ranges. With 50 percent voids between the containers of salt, twice as much total volume is needed. Clearly, the advantages of phase-change materials decline as the storage temperature range increases.

But the costs of these salts can often demolish the best laid plans of enthusiastic designers. Off the shelf, Glaubers salt costs little more than 2 cents a pound. But preparing and putting it in a container can run the costs up. It is unlikely that Glaubers salt will ever be installed in an active solar heat storage system for less than 20 cents a pound. The other salts can cost significantly more.

INSULATION

Every storage system—whether of water, rock, or phase-change material—requires a massive amount of insulation. The higher its average temperature and the colder the surroundings, the more insulation required. For low temperature (below 120°F) storage units inside the house, at least 6 inches of fiberglass insulation (or its equivalent) is the norm. The same unit in the basement needs 8 inches of fiberglass or more. And if located outside, it must be shielded from the wind and insulated even more heavily. The ground can provide insulation if the water table is low. But be careful—even a small amount of moisture movement through the soil will ruin its insulating value.

All ducts or pipes should be just as well insulated as the storage unit. Heat loss from the ducts or pipes can be further reduced by putting the collector close to the storage. The shorter the ducting or piping, the lower the total heat loss. And you'll save on construction and operating costs too.

STORAGE SIZE

The higher the temperature a storage medium can attain, the smaller the storage bin or tank needs to be. For example, 1000 pounds of water (about 120 gallons or 16 cubic feet) can store 20,000 Btu as its temperature increases from 80°F to 100°F, and 40,000 Btu from 80°F to 120°F. It takes almost 5000 pounds of rock (or 40 cubic feet, assuming 30 percent voids) to store the same amounts of heat over the same temperature rises.

Offhand, you might be tempted to design for the highest storage temperatures possible in order to keep the storage size down. But the storage temperature is linked to those of the collector and the distribution system. If the average stor-

age temperature is 120°F, for instance, the heat transfer fluid will not begin to circulate until the collector reaches 135°F. And collector efficiency plummets as the temperature of its absorber rises. A collector operating at 90°F may collect *twice* as much heat per square foot as one operating at 140°F. On the other hand, the storage must be hot enough to feed your baseboard radiators, fan coil units, or radiant panels. For example, a fan coil unit that delivers 120°F air cannot use storage tank temperatures of 100°F without an auxiliary boost in temperature. In general, the upper limit on the storage temperature is determined by the collector performance and the lower limit by the method of heat distribution. You can increase the possible range of storage temperatures and keep the storage size at a minimum by using collectors that are efficient at high tempertures and heat distribution systems that operate at low temperatures.

It's a good idea to allow some flexibiltity in your initial designs so that you can alter the heat storage capacity after some experience under real operating conditions. For example, an oversized concrete water tank can be filled to various levels until the best overall system performance is attained. If you're not too sure of your calculations, the storage should be oversized rather than undersized—to keep its average temperature low.

The capacity of a heat storage unit is often described as the number of sunless days it can keep the house warm. But this approach can be misleading. A system that provides heat for two sunless days in April is much smaller than one that can do so in January. It's better to describe the heat storage capacity as the number of degree days of heating demand that a system can provide in the absence of sulight. For example, a 1200-square-foot house in Minneapolis loses about 10,000 Btu per degree day. In the basement, a tank with 15,000 pounds of water (about 2000 gallons) stores 600,000 Btu as its temperature rises from 80°F to 120°F. Assuming the heating system can use water at 80°F, this is enough heat to carry the house through 60 degree days (or through one full sunless day when the average outdoor temperature is 5°F).

Estimating Storage Size

The following procedure helps you calculate the volume of water or rocks needed to store all the solar heat coming from a collector on an average sunny day. It assumes that the collector performance and size have already been determined according to procedures described earlier.

First you need to determine the maximum storage temperature to be expected. This is 5°F less than the maximum collector operating temperature—considered earlier in "Estimating Collector Performance." The temperature range of the storage medium is the maximum storage temperature minus the lowest temperature that the heat distribution system can use. For example, if the collector can operate at 140°F, the maximum storage temperature is 135°F; and if the heating system can use 85°F, the temperature range is $(135 - 85) = 50°F$.

Next, determine the amount of heat you can store in a cubic foot of the storage medium over this temperature range. This amount is the specific heat of the medium times the density of the medium times the temperature range. For example, a cubic foot of water can store

$$1.0 \ Btu/(lb°F)(62.4 \ lb/ft^3)(50°F) = 3120 \ Btu/ft^3$$

over a 50°F temperature range, or 417 Btu per gallon. If the collector gathers 1000 Btu/ft² on an average sunny day in winter, you need $(1000 \div 417) = 2.4$ gallons of water heat storage per square foot of collector. For the collector on the Boston home, with an area of 277.2 ft², that's 665 gallons of water at the very minimum.

We recall from the Boston example that there are 1088 degree days in January, or 35 per day. The house loses 9500 Btu per degree day, or $(35)(9500) = 332,500$ Btu on an average January day. But the 666 gallons of water can store only 278,055 Btu over a 50°F temperature rise. Therefore, the storage volume must be increased to 796 gallons, or 2.9 gallons per square foot of collector, to satisfy the storage needs of a single January day.

If rock were the storage medium, even more volume would be necessary. A cubic foot of solid rock weighs about 170 pounds and has a specific heat of 0.21 Btu/(lb°F), so it can store

$$0.21 \ Btu/(lb°F)(170 \ lb/ft^3 \ (50°F)$$
$$= 1785 \ Btu/ft^3$$

over the same 50°F temperature range. To store the 1000 Btu from a single square foot of collector, you need $(1000 \div 1785) = 0.56$ cubic feet or 95 pounds of rock. To store the 332,500 Btu required for an average January day, you need $(332,500 \div 1785) = 186$ cubic feet of solid rock.

The total volume occupied by the heat storage container must include void spaces in the storage medium to let the air pass. If there are 30 percent voids between the rocks for example, this Boston home would need a 266-cubic-foot storage bin for the rocks. Or if containerized water with 50 percent voids were used, the 796 gallons or 106 cubic feet of water would occupy 212 cubic feet of house volume.

Generally, the storage should be large enough to supply a home with enough heat for at least one average January day. In Minneapolis, there are about 1600 degree days each January, so the storage unit in our Minneapolis example should be designed to supply at least 52 degree days of heating demand—or 520,000 Btu. Depending upon available funds, the storage can be even larger. Or you can sink your money into better collectors that are efficient at higher temperatures. Solar heat can then be stored at high temperatures—increasing the effective storage capacity of a tank or bin.

At the very least, the storage should be large enough to absorb all the solar heat coming from the collectors in a single day. If you can be satisfied with 60-percent solar heating or less, the simplified method described in "Estimating Collector Size" will be useful. First you size the collector according to the method provided earlier under "Estimating the Collector Size." Then determine the volume of storage medium required per square foot of collector. Multiplying this volume by the total collector area gives a reasonable "first-cut" estimate of storage size. This estimate should be close enough for preliminary design work. If this storage volume fails to meet the heat demand for an average January day, revise your estimate upward until it does.

To get more than 60-percent solar heating, it helps to know the normal sequence of sunny and cloudy days in your area. If sunny days followed cloudy days one after the other, you would only have to size the collector and storage for one sunny day and the following cloudy day. Almost 100 percent of the heating demand could then be provided if the system were designed for the coldest two-day period. If the normal sequence were one sunny day followed by two cloudy days, both collector and storage size would have to be doubled to achieve the same percentage. At the Blue Hills weather station near Boston, for example, about 80 percent of the sunless periods are two days long or less. A collector and storage system that could carry a house in Blue Hills through two cloudy days of

the coldest weather will supply more than 80 percent of the home's heating needs. But the wide variation in weather patterns at a single location makes such a practice little more than educated guesswork. And this kind of weather data is hard to obtain.

HEAT DISTRIBUTION

An active solar heating system usually requires another sub-system to distribute the heat to the rooms. With integrated solar heating methods such as mass walls and direct-gain windows, the solar heat is absorbed directly in the fabric of the house and heat distribution comes naturally. But an active system usually needs more heat exchangers, piping, ducts, pumps, fans, and blowers to get the heat inside the house. And there must still be some provision for backup heating in the event of bad weather.

The heat distribution system should be designed to use temperatures as low as 75°F to 80°F. If low temperatures can be used, more solar energy can be stored and the collector efficiency increases dramatically. In general, warm air heating systems use temperatures from 80°F to 130°F, while hot water radiant heating systems require temperatures from 90°F to 160°F. Steam heating is rarely combined with a solar array since it needs temperatures over 212°F. so incorporating solar heating into an existing house equipped with steam heating will require a completely separate heat distribution system.

Many designers of new homes opt for forced warm air systems or radiant heating systems. By using larger volumes of air or oversized panels (such as concrete floors), these solar heating systems can operate at lower temperatures. Radiant slabs take time before the room is comfortable and usually require slightly higher storage temperatures. But because their radiant heat warms occupants directly, the air temperature can be kept 2 to 3°F lower, reducing the home's heat loss.

AUXILIARY HEATING

Even a system with a very large storage capacity will encounter times when the heat is used up. So the house must have an auxiliary heat source. This is a major reason why solar heating has not yet met with widespread acclaim—you still have to buy the conventional heating system.

The severe consequences of a single sustained period of very cold, cloudy weather are enough to justify a full-sized conventional heating system as backup. Small homes in rural areas can probably get by with wood stoves. If the climate is never too severe, as in Florida and most of California, a few small electric heaters may do the trick. But most houses will require a full-sized gas, oil, or electric heating system. Solar energy is a means of decreasing our consumption of fossil fuels—not a complete substitute. Energy conservation in building construction, the first step to a well-designed solar home, lowers the building's peak heat load, which means you can buy a smaller, less-expensive auxiliary heating system.

If the auxiliary system won't be needed very often, you might well consider electric heating. But remember that 10,000 to 13,000 Btu are burned at the power plant to produce 1 kilowatt-hour of electricity—the equivalent of only 3400 Btu in your house. At an efficiency of 65 percent, an oil furnace burns only 5400 Btu to achieve the same result. And electric heating can be very expensive, although the first cost of electric heaters is cheaper than a gas or oil furnace.

The auxiliary heater should not be used to heat the storage tank or bin because the collector will operate at a higher temperture and lower efficiency. And there will be costly heat losses from the storage container if an auxiliary sytem provides continuously higher storage temperatures. The heat lost from the storage container is already 5 to 20 percent of the solar heat collected.

The heat pump has served as a combination backup and booster in a number of solar energy systems. It is basically a refrigeration device working in reverse. The heat pump takes heat from one location *(the heat source)* and delivers it to another *(the heat sink)*. The heat source is cooled in the process and the heat sink is warmed.

A heat pump can deliver about three times the energy required for its operation. For every 2 Btu which a heat pump takes from a source, it needs the equivalent of 1 Btu of electricity for its operation. It delivers all 3 Btu to the heat sink. Thus, its *Coefficient of Performance* (COP), or the ratio of the heat energy delivered to the energy required for operation, is 3. Typically, this coefficient ranges from 2.5 to 3 for good heat pumps. By contrast, electric resistance heating has a COP equal to 1, because it delivers 1 Btu of heat for every 1 Btu of electricity expended.

When heat pumps are used in conjunction with solar heating sytems, the stored heat is useful over a wider temperature range. Without heat pumps, a forced warm air system would use storage tempertures from 80°F to 130°F and a hot water radiant system would use 90°F to 160°F. But with a heat pump, both systems can use 40°F storage temperatures! The heat pump takes low grade heat from storage and delivers it at a higher temperature to the heat distribution system. This increased temperature range results in an increased heat storage capacity and markedly enhanced system efficiency. The extra Btu that a cool collector can gather each year often justify the added cost of installing a heat pump.

But heat pumps require electricity for operation. About 10,000 to 13,000 Btu are burned at the power plant when a heat pump uses 1 kilowatt-hour of electricity (or 3400 Btu) to deliver a total of 10,200 Btu. So, including losses at the power plant and in electrical transmission, the real Coefficient of Performance is closer to 1 than 3. And electricity *is* expensive. High electricity bills have been a major shortcoming of past solar heating systems that relied on heat pumps.

Heat Pump Principles

A heat pump is a mechanical device that transfers heat from one medium to another, thereby cooling the first and warming the second. It can be used to heat or cool a body of air or a tank of water, or even the earth. The cooled medium is called the "heat source" and the warmed medium is the "heat sink." A household refrigerator is a heat pump that takes heat from the food compartments (the heat source) and dumps it in the kitchen air (the heat sink).

The heat pump transfers heat against the grain—from cool areas to warm. This sleight-of-hand is accomplished by circulating a heat transfer fluid or "refrigerant" (such as the Freon commonly used in household refrigerators) between the source and sink and inducing this fluid to evaporate and condense. Heat is absorbed from the source when the heat transfer fluid evaporates there. The vapor is then compressed and pumped through a heat exchanger in the sink, where it condenses—releasing its latent heat. The condensed liquid returns to the heat exchanger in the source through and expansion value, which maintains the pressure difference created by the compressor.

The packaged, self-contained heat pump used in residential applications generally reverses the direction of the refrigerant flow to change from heating to cooling or vice-versa. A four-way value reverses the direction of flow through the compressor so that high pressure vapor condenses inside the conditioned space when heating is needed and low pressure liquid evaporates inside when cooling is desired. Heat pumps are classified according to the heat source and sink, the fluid used in each, and the operating cycle. The heat pump shown here is a water-to-water pump with reversible refrigerant flow. A household refrigerator is an air-to-air heat pump with a fixed refrigerant flow. Ground-coupled heat pumps are usually water-to-air, but are occasionally water-to-water.

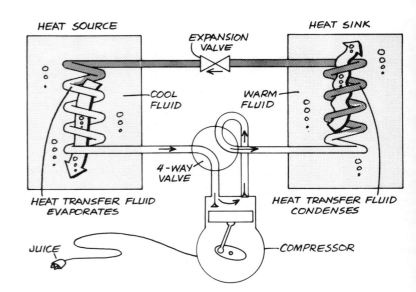

Coefficient of Performance

A heat pump uses electrical energy to manipulate heat transfer from source to sink. The heat deposited in the sink is a combination of the heat generated by compressing the refrigerant (which requires electrical power) and the latent heat released by the condensing vapor. The heat removed from the heat source is the latent heat of evaporation. The effectiveness of a heat pump is indicated by its Coefficient of Performance, or COP, which equals heat energy deposited (or removed) divided by electrical energy consumed.

The electrical energy required to run the compressor (in kwh) can be converted to Btu by multiplying by 3413 Btu/kwh. Because the heat of compression is part of the heat deposited in the sink, the COP of a heat pump used for heating is usually greater than the COP of the same heat pump used for cooling.

16
Photovoltaics:
Electricity from the Sun

Of all the benefits the sun can give us, potentially the most far reaching is direct generation of electrical power. Photovoltaic (PV) cells can generate electricity from sunlight. Like sunlight, electricity is an essentially benevolent form of energy: silent, invisible, quick, nonpolluting, far reaching. PV-powered vehicles and aircraft are now where Henry Ford and the Wright brothers were with their fossil-fueled inventions less than a century ago. Imagine a future where pollution-free cars and airplanes whisk us silently from one place to another: where trucks, buses, machines, and factories no longer intrude upon the natural landscape and atmosphere.

The use of solar cells to generate electricity is a very recent achievement. Although the photovoltaic phenomena was first discovered in the 19th centrury, it was not until the 1950s that scientists built the first working solar cells. Because of the high costs involved, solar cells were used initially in military and research projects and where the cost of obtaining conventional power was too expensive. During the 1960s and 1970s, solar cells were used to provide power in remote installations, such as electronic relay stations, irrigation pump facilites, and navigational buoys, as well as for homes and other installations that were not tied to conventional power lines. Other applicatons were developed by using solar cells to charge storage batteries, providing a steady source of power for devices such as marine radios, lights, and recorders.

High cost is still a serious disadvantage of PV systems. Although considerably cheaper than the $5,000 per watt cost of 25 years ago, PV power is still more expensive than conventional utility power.

SUNLIGHT TO ELECTRICITY

Photovoltaic or solar cells generate electricity when exposed to sunlight. Although the amount of electricity produced by a single solar cell is small, a group or array of cells can generate a considerable amount—almost 1/2 kilowatt per square foot of cell surface. A sufficiently large array—555 square feet for a small residential installation—can generate enough electricity to meet a substantial percent of the needs of a single-family home.

In practice, electricity generated by PV arrays can be used as direct current (dc), stored in batteries for subsequent use, or changed to alternating current (ac) for immediate use. With a good-quality power inverter (a transformer-like device that converts dc to ac), surplus electricity can be fed into the local power grid to be credited against later power drawn from that system when the photovoltaic array is not gen-

138

erating electricity. Local power companies are required by federal law to accept and pay for such customer generated power. This eliminates the need for costly battery storage.

When bundles of solar energy, called photons, penetrate two-layered solar cells, they knock loose electrons, transferring energy to them. These loose electrons move to one side of the cell, creating a negative charge. On the other side, a deficit of electrons creates a positive charge. As they move about, these loose electrons are quickly caught up in an electrical field, forming a weak electric current at the junction of the two layers.

To generate electric current in this manner, solar cells use a semiconductor, typically two layers of silicon. The negative layer on the top facing the sun is treated or "doped" with phosphorous to create an excess of electrons; the positive layer on the bottom of the cell is treated with boron to create vacancies or "holes" for new electrons to fill. As photon energy is absorbed by the negative layer, millions of excess electrons are captured by the electrical field at the junction between the two layers. The voltage difference between them pushes the electrons through a wire grid on the front of the cell, which is connected in turn to the wire grid on the back of the next cell. As current flows through a series of cells, its voltage continues to build.

The *cells* are sandwiched together between a substrate and a superstrate to form a *module*. The aluminum-framed modules are connected together to form *panels* that are installed in an *array*. The ultimate current and voltage produced by the array depends on how the modules and panels are wired together.

Most PV semiconductors are made of crystalline silicon in an expensive manufacturing process. Among alternative semiconductor materials less expensive to produce, amorphous silicon offers much of the same capability as crystalline silicon. Amorphous silicon actually absorbs visible light better than the crystalline form, but it is less efficient—about 3 percent compared with 12 to 16 percent for crystalline in converting light to electricity. Nevertheless, the lower material and fabrication cost make

amorphous silicon a prime candidate to replace crystalline silicon as a low-cost source of PV power in the future.

POWER REQUIREMENTS

A typical residential PV installation consists of an array of solar modules, an inverter to change solar-cell direct current to alternating current and, where local utility power is unavailable, a bank of batteries to store excess electricity.

In designing a system, you must take into account factors such as geographical locations, availability of local utility power and how much electricity you need.

Geographic location affects the amount of potential power available from a system, because the percentage of sunshine available varies greatly in different sections of the country. A home in the southwestern U.S. can count on a much higher percentage of daily sunshine than one in the northeast or coastal northwest. A residential PV system in Arizona for example, will generate almost twice as much electricity as a similar system in New Hampshire. To determine the amount of sunshine available in your location you can refer to the *Climatic Atlas of the United States* which lists percentages of possible sunshine by geographic area. The less sunlight available, the larger your PV array must be to meet given power during needs.

If you live in an area where local utility power is unavailable, you must design your system to store power for use when there is not enough sunlight for PV operation. Storage requires a bank of batteries with sufficient capacity to provide power at night and during cloudy weather.

AN AVERAGE HOME

The table that follows lists the example requirements for an average home. But how much of this load could be met by a PV system?

Let's say that the house with the electrical demand we just calculated is located in Phila-

Estimating Array Size

A prime consideration in your design is the amount of power your home needs and the pattern of daily usage. These factors determine the size of the system and its components. Chances are that the power needs of appliances and lighting in your home will exceed the capability of most moderate-size residential installations. The solution is energy conservation—reducing and planning your needs.

First determine your need for power by making a list of your appliances and other electrical devices. Record their power requirements in watts, how many hours each is used on a weekly

basis, and the percentage of time the appliance is normally running. With this information, calculate your initial average overall requirements per day. For example, to calculate how much power your refrigerator will require on an average day, figure the weekly average and divide by 7. Its rated power requirement is 400 watts. Since the refrigerator is used 24 hours a day (168 hours in 7 days) and it runs about one half (50%) of the time, multiply these figures together to get its energy requirement:

$$400\,watts(168\,hours)(.50) = 33.6\,kwh\,per\,week / 7\,days\,per\,week = 4.8\,kwh/day$$

delphia (40° north latitude). There is room on its 500-square-foot roof for the PV array. The roof is pitched at a 40° slope. First, we must calculate how much solar radiation is available every month. For example, in January, the average daily insolation (from the "Clear Day Insolation" tables in the appendix) on a 40° slope is 1810 Btu/(ft^2 day). From the "Mean Percentage of Possible Sunshine" map in the appendix, we see that Philadelphia receives only 50 percent of the possible sunshine in January. Multiplying 0.50 by 1810 Btu/(ft^2 day) and 500 square feet, we calculate that the roof would receive 452,500 Btu/day of sunshine. Since there are 3412 Btu in a kwh, that is 133 kwh/day.

But not all that energy can be converted into electricity. If the solar cells only have an efficiency of 0.10, then the array only produces (133)(0.10) or 13.3 kwh/day. The inverter and other balance of system parts lose another 15 percent in conversion of the dc power to ac, so the system output is further reduced by (13.3)(0.85) to 11.3 kwh/day.

If our daily average demand is 19.4 kwh, the PV array could provide (11.278/19.4)(100) or 58 percent of the electricity used. Since not all the electricity is used during daylight hours, some of the demand would have to be met by

the power company, but some of the power produced by the PV array would be stored in batteries for later use, or flow back into the grid to be credited against the power bought.

The second table lists the average radiation that strikes the roof each month, the percent possible sunshine for that month, the total daily insolation on the roof, how many kwh a day the system produces after subtracting for cell efficiency (0.10) and balance of system efficiency (0.85). The last column is the percent of the daily demand supplied by the array.

Remember that this only gives you an estimate of *what* the array may produce, and not a guarantee of *how much* energy you'll collect. That depends on your system components and local climate.

If you have unlimited roof area and would like to size the array based on the demand, you can find a range of areas that will help you decide. With the monthly insolation values and percent possible sunshine, you can find the maximum and minimum array you need.

In our example, December gets the least sun with only 815 Btu/(ft^2 day) from (1634 Btu/(ft^2 day))(0.5). In August, the roof gets the most sun with 1400 Btu/(ft^2 day) from (2258 Btu/(ft^2 day))(0.62). Converting the solar gains to kwh,

EXAMPLE POWER REQUIREMENT

Appliance	Power Required (Kw)	Hours/ Week	Percent Running Time	Average Kwh/Week
Refrigerator	400	168	50	33.6
Dish washer	1000	5	100	5.0
Clothes washer	600	2	100	1.2
Clothes dryer	4500	2	100	9.0
Stove	3500	7	100	24.5
Water heater	3000	168	10	50.4
Oil furnace pump	250	168	20	8.4
Toaster	1000	0.07	100	0.07
Record/tape player	100	7	100	0.70
Television set	200	7	100	1.40
Radio	40	14	100	0.56
Electric saw	450	2	100	0.90

Total Kwh per week 135.73
Average Kwh per day 19.39

EXAMPLE GAIN FROM 500 SQ FT ROOF ARRAY

Month	(1) Surface Daily Total Insolation Btu / (ft^2 day)	(2) Percent Possible Sunshine	(3) Total Daily Insolation (Kwh/day) (1) x (2) x 500 / 3412	(4) System-Produced Energy (Kwh/day) (3) x 0.1 x 0.85	Percent Supplied (4) / 19.4 x 100
January	1810	0.50	133	11.3	58%
February	2162	0.60	190	16.2	83
March	2330	0.55	188	16.0	82
April	2320	0.55	187	15.9	82
May	2264	0.60	199	16.9	87
June	2224	0.62	202	17.2	89
July	2230	0.60	196	16.7	86
August	2258	0.62	205	17.4	90
September	2228	0.60	196	16.6	86
October	2060	0.60	181	15.4	79
November	1778	0.50	130	11.1	57
December	1634	0.50	120	10.2	52
				Average	78%

SIZING THE ARRAY TO DEMAND

Month	System Size (ft^2) Based on Monthly Insolation	Energy Produced with 953 ft^2 Array		Energy Produced with 557 ft^2 Array	
		Kwh/Day	% Excess	Kwh/Day	% Deficit
January	860	21.5	+11%	12.6	-35%
February	600	30.8	59	18.0	7
March	608	30.4	57	17.8	8
April	610	30.3	56	17.7	9
May	573	32.3	66	18.8	3
June	555	32.7	69	19.1	1
July	582	31.8	64	18.6	4
August	556	33.2	71	19.4	0
September	583	31.7	64	18.5	4
October	630	29.3	51	17.2	12
November	876	21.1	9	12.3	36
December	953	19.4	0	11.3	42
		Average	+48%	Average	-13%

the maximum gain is 0.41 kwh/(ft^2 day) and the minimum is 0.24 Btu/(ft^2 day).

The areas needed to supply the demand can be found by: area (ft^2) = daily demand/[(solar gain) (cell efficiency) (balance of system efficiency)].

In our example:

$$\text{maximum area} = 19.4/[0.24(0.10)(0.85)]$$
$$= 953 \text{ ft}^2$$
$$\text{minimum area} = 19.4/[0.41(0.10)(0.85)]$$
$$= 557 \text{ ft}^2$$

Depending on how much money you want to spend on the system, and how much you want to invest in battery storage (for the excess) or how much power you want to buy from the electric company (from the deficit), your array should be between 557 and 950 square feet. The next table shows how much excess or deficit energy those two PV array sizes will produce each month. The first column shows how big the array should be based on the average energy produced per day that month. The second shows the energy produced per month if the system were sized at the maximum and the third how much energy would be produced if the system were sized to the minimum. Unless your power company is paying top dollar for the electricity it buys from you, you'd be better off sizing the collector toward the minimum side to save on the high first cost of the system.

Your initial overall requirement for power is likely to be substantial, well beyond the capacity of a residential PV system. To reduce the amount of power needed to a more practical level, conserve energy first. Reduce the number of electrical appliances. Use them less. When the appliance requires a large amount of power, e.g., a refrigerator, hot water heater, or clothes dryer, install more efficient units or replace them with non-electrical devices, such as a solar water heater, or a clothes line for drying. Your goal is to reduce the gap between the amount of power needed and the amount your PV system will generate.

Daily residential power use usually shows peak consumption at meal times and in the evening with an additional low level of steady usage 24 hours a day. Your system must be able to provide adequate power at these times plus when it is dark or cloudy. It is important to note that electric motors, such as those used in refrigerators and washing machines, require a large surge of electricity when they start. Your system has to meet these extroardinary peak needs also.

SUPPLEMENTAL POWER

To meet all of your power needs, you will probably have to supplement your PV power with either local utility power or battery storage.

Local utility power enters a residential PV system through an inverter, which also converts dc power from the solar cells to ac. As ac power from the utility is used, it is metered in the usual way to determine the number of kilowatt hours used. When power from the PV system reaches a sufficient level, the inverter cuts off utility power. The inverter also directs excess PV-system-generated power into the local utility lines, in effect running the meter backward. Because power fed into a utility power line must meet certain standards of electrical quality, inverters must be properly matched to the utility system.

Battery storage can be used to meet both 24-hour and peak-load needs. (Batteries are essential where no local utility power is available.) Charged by electricity from solar cells during hours of sunlight, batteries store power, making it available for use by appliances and lights when needed. Where appliances and lights can be operated on dc, no inverter is necessary. However, most appliances use ac, and so need an inverter to convert dc to ac current.

The number and overall voltage and current output of the batteries depend on power needs as well as on the amount of power provided by the PV system. High voltages (32 to 38 volts) are much more efficient than low voltages and are necessary to meet most modern residential power requirements. Depending on geographic location, a PV system may need to rely on battery power for two or three weeks of cloudy weather at a time. A gasoline generator can be used to charge batteries during sunless periods, reducing the number of batteries required.

POWER INVERTERS

Power inverters are essential in ac systems to convert solar-cell generated dc to residential-system ac. They are also essential as a go-between with the utility line.

The inverter is the control center of the PV system. It turns on the system when sufficient power is available from solar cells. It turns it off when power drops below a set level. The inverter also determines the quality of the alternating current waveform, which powers residential appliances. Devices such as stereo turntables require a high-quality waveform. When a PV system is tied to local utility lines, the inverter also must provide an ac waveform compatible with the utility line. Only a high-quality inverter—solid state or synchronous—can meet these requirements.

RESIDENTIAL INSTALLATIONS

A typical residential PV system has 550- to 860-square-feet of roof-mounted, interconnected PV modules. Such an array can produce 5 to 8 kilowatts at maximum voltages of 160 to 200 volts. At efficiencies of 85 to 90 percent, an inverter will yield a peak rate between 6400 and 9000 kwh of power in regions with moderate sunshine (for example, the Northeast) and an additional 80 percent—11,500 kwh to 16,200—in the Southwest.

While such a system may not provide 100 percent of a family's electrical needs, it can provide 50 to 90 percent, depending on location and amount of sunshine.

Residential PV panels can be mounted in several ways; stand-off, direct, rack, or integral. A stand-off mount places PV panels several inches above the roof, which allows air to circulate behind the cells to cool them and increase their efficiency. These panels are attached to mounting rails which, in turn, are attached to the roof rafters.

Direct-mount arrays are those attached directly to roof sheathing, replacing the roofing material. Like shingles, these special PV modules overlap each other, forming an effective seal against water and wind. However, operating temperatures will be higher and efficiency lower because of the lack of air circulation around the back of the cells to cool them. For this reason, direct-mount arrays are few and far between.

Rack-mounted PV panels are used on flat roofs to position PV cells at the proper angle to the sun. Although more complicated to build than a stand-off mount, this arrangement provides excellent air circulation, increasing system efficiency.

Integral-mounted PV systems replace ordinary roof sheathing and shingles. PV panels are attached directly to rafters, the space between them sealed with gaskets. Other edges are sealed with silicone sealant or held down by aluminum battens. Attic ventilation keeps backside temperatures at efficiently cool levels.

Integral mounts can produce hot air for space heating or solar domestic hot water by directing the air used to cool the back of the cells to the load or to a storage container.

In designing a mounting system, there are several factors to consider. Panels must be installed so that they are easily accessible. For example, debris must be brushed away from time to time and accumulated dirt washed off cell surfaces. Stand-off and rack panels must be easy to remove from their mountings for repair or module replacement. Panels installed intergrally must be water-tight and panel backs accessible for repair and cooling. Whenever possible PV panels should be mounted to allow air to circulate behind them to cool the underside of the cells.

The appearance of the PV array is also important—especially roof-mounted arrays that are clearly visible. Rectangular and square cells and dark anodized metal frames blend in better with most roofs than modules of round cells and polished aluminum frames.

Epilogue

Solar energy, in the last analysis, has always been the basis not only of civilization, but of life; from the primeval sun-basking plankton to modern man harvesting his fields and burning coal and oil beneath his boilers, solar energy has provided the ultimate moving force. But its direct utilization at a higher level of technology is a new phenomenon, and rich with new potentialities at this stage of human affairs.

Peter van Dresser,
Landscape, Spring 1956

There are still many obstacles blocking the widespread use of the sun's energy for heating and cooling. Most of these problems are *nontechnical* in nature, having to do with solar energy's impact upon and acceptance by society as a whole. Whenever a new building method bursts upon the scene, financial institutions and the building trades are understandably conservative until that method has proved itself. But with the long-range depletion of cheap fossil fuels and the rising energy needs of our developing world, the rapid development of this heretofore neglected power source is inevitable.

FINANCIAL CONSTRAINTS

The greatest barrier to the immediate home use of solar energy is the high initial cost. Depending upon size and complexity, a solar heating system could add 2–10 percent to the building cost of a new house. A system fitted to an existing house costs more. Financing such an expenditure is particularly difficult during periods when costs are burdensome, interest rates high, and mortgage money difficult to obtain. Financing is one of the principal reasons people decide against using solar energy.

Part of the problem, of course, is that solar energy is still not a major established alternative to conventional heating systems. Banks are reluctant to fund an expensive addition that they consider unlikely to pay back. As more solar heating systems come into general use, however, and bear out the claims of lowered heating costs, loans for these systems are becoming more readily available.

Compounding the financing difficulties is the fact that an auxiliary heating system must be provided, even in solar-heated homes. People prefer complete heating systems rather than systems that provide only 50–90 percent of their heating needs. But 100-percent solar heating systems are usually far too large to be practical, so the additional expense of a conventional heating system must also be borne.

If long-range predictions are true, and sources of conventional fuels dry up over time, arguments against a big cash outlay will lose their

Life-Cycle Costing

Life-cycle costing is an estimating method that includes the future costs of energy consumption, maintenance and repair in the economic comparison of several alternatives. These future costs can make an initially cheaper system costlier over the life cycle of the system. Life-cycle costing methods make such costs visible at the outset, and they include the economic impact of interest rates and inflation. They are ideally suited for comparing the costs of solar heating with those of conventional methods.

In order to obtain consistent cost comparisons among several alternatives, all the costs of each system (over a selected "life cycle") are reduced to total costs over a unit of time, usually the first year. Future savings such as lower fuel costs are discounted to "present-value" dollars, which is the amount of money that, if invested today, would grow to the value of the savings in the intervening years. And if the annual operating and maintenance expenses can be predicted to grow at some steady inflation rate, the present-value total of those expenditures over the life cycle of the system (P_e) can be calculated using the following equation:

$$P_e = A(R)(R^n - 1)/(R - 1)$$

where $R = (1 + g)/(1 + i)$ and i and g are the fractional rates of interest and inflation. In this equation, the current annual expense (A) is multiplied by a factor which accounts for the number of years in the life cycle (n) and the rate at which the annual expense (A) is expected to increase.

Example: Assume that the retail cost of heating oil is $1.00 per gallon and that it will increase 5 percent per year. What is the present value of the expenditures for one gallon of oil each year for the next 30 years?

Solution: Assuming an annual interest rate of 10 percent, the ratio R equals 1.05/1.12, or 0.9375. Applying the equation, we find the present value of 30 gallons of oil expended over the next 30 years is $12.84:

$$P_e = \$1.00(0.9375)(0.9375 - 1)/(0.9375 - 1)$$
$$= (\$1.00)(12.84) = \$12.84$$

To get the life-cycle costs of a system, the purchase and installation prices are added to the present value of the total operating and maintenance costs.

For example, an owner-builder might want to compare the life-cycle costs of insulated 2x4 stud walls to the cost of insulated 2x6 stud walls. He estimates that the 2x4 walls will cost $7140 to build but will lose 48.4 million Btu per year; the 2x6 walls will cost $7860 and lose 34.8 million Btu. Assuming that a gallon of heating oil produces 100,000 Btu of useful heat, the house will require 484 or 348 gallons per year, depending upon the wall construction. Over a 30-year life cycle, the operating costs will be ($12.84)(484) = $6215 for the 2x4 walls and ($12.84)(348) = $4468 for the 2x6 walls, in present-value dollars. Maintenance expenses are equal for the two alternatives and hence are ignored. Adding operating and installation costs, the owner-builder finds that the 30-year life-cycle costs of the two alternatives are:

2x4 walls = $7140 + $6215 = $13,355
2x6 walls = $7860 + $4468 = $12,328

Life-cycle costing suggests that the initially more expensive 2x6 wall is actually more than $1000 cheaper over time. Similar costing can be used to compare the costs of solar heating and of conventional systems.

clout and the *availability* of fuel will become the real issue. People who find themselves without fuel will decide that the shortage is reason enough for using solar energy, that the initial costs of the system are less important.

But home-financing plans can encourage such an investment even now because lower heating bills *over the lifetime* of the system make it a sound buy. All too frequently, financial institutions disregard the ever-increasing operating costs of a conventionally heated home and focus upon the large initial costs of a solar heating system. Some lending institutions, however, are using life-cycle costing methods, which compare the higher initial costs of solar to lower operating and maintenance costs. These methods emphasize the lowest total monthly home-owning costs (mortgage payments *plus* utilities). Lenders should allow higher monthly mortgage payments if monthly energy costs are lower. Most progressive lending institutions are doing just that.

SYSTEM RELIABILITY

A major difficulty in the custom design and manufacture of active solar heating systems for particular sites is the necessary combination of low cost, good performance, and durability. The building designer must have a thorough understanding of the principles of solar energy and of the pitfalls discovered in the past. Even then, many things can go wrong with such complex systems, and many architectural and engineering firms hesitate to invest extra time and money in custom designs. Most active systems today, however, are designed by the supplier of the equipment, leaving the building designer with the responsiblity only to select the best system from the most reliable supplier.

One of the most appealing aspects of solar heating has been that custom design and on-site construction often seems a cheaper alternative than buying manufactured collectors. However,

this turns out usually not to be the case as construction costs rise and prices of manufactured systems drop. (The exception would be systems built by the do-it-yourselfer, or the owner-builder, whose time and labor are not usually counted in the cost.) There are now hundreds of excellent solar products, and the competition is fierce. Resulting price reductions are inevitable in the long run, but they will come slowly.

Passive solar systems are often cheaper, more efficient, and more reliable than active systems, and are usually more appropriate for custom design and on-site construction of new houses. Here, too, there are now many products to choose from.

SOLAR ENERGY AND THE CONSTRUCTION INDUSTRY

The housing industry and the laws that regulate it have a record of slow adaptation to change. The industry is very fragmented, with thousands of builders, and 90 percent of all work is done by companies who build fewer than 100 units per year. The profit margin is small, and innovation is a risk that few builders will take.

But the fragmented nature of the construction industry is essential to the localized industries that have sprung up around solar energy. Even if a few large manufacturers achieve low-cost solar collectors, interstate transportation costs will remain relatively high, adding one or two dollars per square foot to collector costs. On-site construction or local fabrication of components will be a viable alternative for years to come.

Some contractors and developers install solar equipment in order to evoke interest in their recent housing developments. And, despite lessening public concern for energy, most builders and contractors are building energy efficient homes, and more and more are including solar water heating and passive solar room heating systems.

147

The New Solar Home Book

GOVERNMENT INCENTIVES

Some local governments still acknowledge the benefits of solar energy in their tax laws. The extra employment stimulated by solar energy, which requires local labor to build and install components, can be a boon to local economies. Annual cash outflows for gas and oil from energy-poor areas like New England amount to billions of dollars that can be saved through energy conservation and the use of solar energy. Also, reductions in pollution levels result from lowered consumption of burning fuels. Local taxes can discourage solar systems, if installation costs mean higher property taxes for the owner. Lowering taxes to encourage the use of solar energy is a desirable goal, and many communities now do not add the value of a solar energy system to the assessed value of a home.

Government incentives and development programs are important to the further development of solar energy. Solar energy must overcome many obstacles, apart from competition with established, government-subsidized energy suppliers such as the nuclear and oil industries. The importance of solar energy to global and national welfare is more than adequate justification for equal promotion of it.

But more than technological innovation and government incentive will be needed to make solar systems a universal reality. In a larger sense, a nation's energy future rests in the personal choices of its people. The consumption practices learned in an age of plentiful and cheap fossil fuels cannot be supported by a solar economy. We can enjoy clean, inexhaustible solar energy much sooner if we insist that our energy-consuming possessions (houses, cars, appliances, etc.) be energy efficient. Also, we should take a cue from the more intimate relationship between humans and their natural world that prevailed in the centuries prior to the availability of cheap fossil energy. People had simple needs that could be supplied by the energy and materials around them. They interacted with their climates to take full advantage of natural heating and cooling. These attitudes of efficiency and harmony with the environment must once again become standard. A new solar age will dawn when we can forego our high-energy ways of life and return to our place in the sun.

148

Appendixes

1

Solar Angles

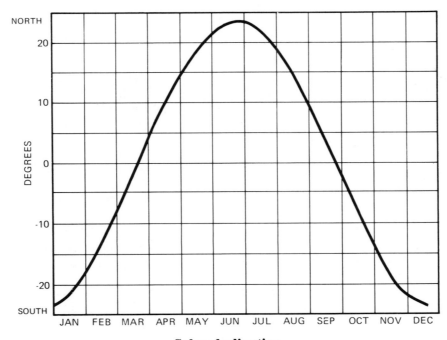

Solar declination

The sun's position in the sky is described by two angular measurements, the solar altitude (represented by the Greek letter theta or θ) and the solar azimuth (represented by the Greek letter phi or φ). As explained earlier in the book, the solar altitude is the angle of the sun above the horizon. The azimuth is its angular deviation from the true south.

The exact calculation of theta or phi depends upon three variables: the latitude (L), the declination (represented by the Greek letter delta or δ), and the hour angle (H). Latitude is the angular distance of the observer north or south of the equator; it can be read from any good map. Solar declination is a measure of how far north or south of the equator the sun has moved.

149

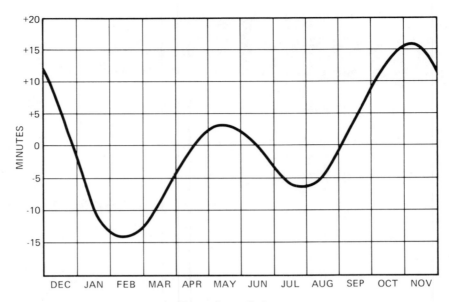

Equation of time

At the summer solstice, $\delta = +23.5°$, while at the winter solstice $\delta = -23.5°$ in the northern hemisphere; at both equinoxes, $\delta = 0°$. This quantity varies from month to month and can be read directly from the first graph shown here.

The hour angle (H) depends on Local Solar Time, which is the time that would be read from a sundial oriented south. Solar time is measured from solar noon, the moment when the sun is highest in the sky. At different times of the year, the lengths of solar days (measured from solar noon to solar noon) are slightly different from days measured by a clock running at a uniform rate. Local solar time is calculated taking this difference into account. There is also a correction if the observer is not on the standard time meridian for his time zone.

To correct local standard time (read from an accurate clock) to local solar time, three steps are necessary:

1) If daylight savings time is in effect, subtract one hour.
2) Determine the longitude of the locality and the longitude of the standard time meridian (75° for Eastern, 90° for Central, 105° for Mountain, 120° for Pacific, 135° for Yukon, 150° for Alaska-Hawaii). Multiply the difference in longitudes by 4 minutes/degree.

If the locality is east of the standard meridian, add the correction minutes; if it is west, subtract them.
3) Add the equation of time (from the second graph shown here) for the date in question. The result is Local Solar Time.

Once you know the Local Solar Time, you can obtain the hour angle (H) from:

$$H = 0.25(\text{number of minutes from solar noon})$$

From the latitude (L), declination (δ), and hour angle (H), the solar altitude (θ) and azimuth (ϕ) follow after a little trigonometry:

$$\sin \theta = \cos L \cos \delta \cos H + \sin L \sin \delta$$
$$\sin \phi = \cos \delta \sin H/\cos \theta$$

As an example, determine the altitude and azimuth of the sun in Abilene, Texas, on December 1, when it is 1:30 p.m. (CST). First you need to calculate the Local Solar Time. It is not daylight savings time, so no correction for that is needed. Looking at a map you see that Abilene is on the 100°W meridian, or 10° west of the standard meridian, 90°W. Subtract the 4(10) = 40 minutes from local time; 1:30 − 0:40 = 12:50 p.m. From the equation of time for December 1, you must *add* about 11 minutes. 12:50 + 0:11 = 1:01 Local Solar Time, or 51 minutes past solar noon. Consequently, the hour angle is H = 0.25(61) or about 15°. The latitude of Abilene is read from the same map: L = 32°, and the declination for December 1 is δ = − 22°.

You have come this far with maps, graphs, and the back of an old envelope, but now you need a scientific calculator or a table of trigonometric functions:

$$\sin θ = \cos(32°)\cos(-22°)\cos(15°)$$
$$+ \sin(32°)\sin(-22°)$$
$$= 0.85(0.93)(0.97) + 0.53(-0.37)$$
$$= 0.76 - 0.20 = 0.56$$

Then θ = arcsin(0.56) = 34.12° above the horizon. Similarly:

$$\sin θ = \cos(-22°)\sin(15°)/\cos(34.12°)$$
$$= (0.93)(0.26)/0.83 = 0.29$$

Then φ = arcsin(0.29) = 16.85° west of true south. At 1:30 p.m. on December 1 in Abilene, Texas, the solar altitude is 34.12° and the azimuth is 16.85° west.

Sun Path Diagrams

In applications where strict accuracy is superfluous, solar angles can be quickly determined with sun path diagrams. In these diagrams, the sun's path across the sky vault is represented by a curve projected onto a horizontal plane (see diagram). The horizon appears as a circle with the observation point at its center. Equally-spaced concentric circles represent the altitude angles (θ) at 10° intervals, and equally spaced radial lines represent the azimuth angles (φ) at the same intervals.

The elliptical curves running horizontally are the projection of the sun's path on the 21st day of each month; they are designated by two Roman numerals for the two months when the sun follows approximately this same path. A grid of vertical curves indicate the hours of the day in Arabic numerals.

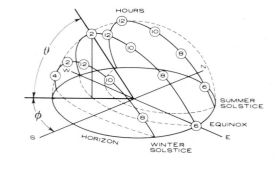

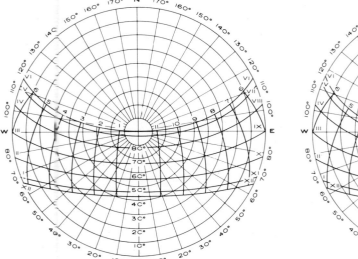

24°N LATITUDE

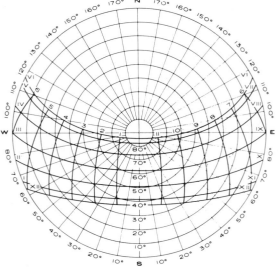

28° N LATITUDE

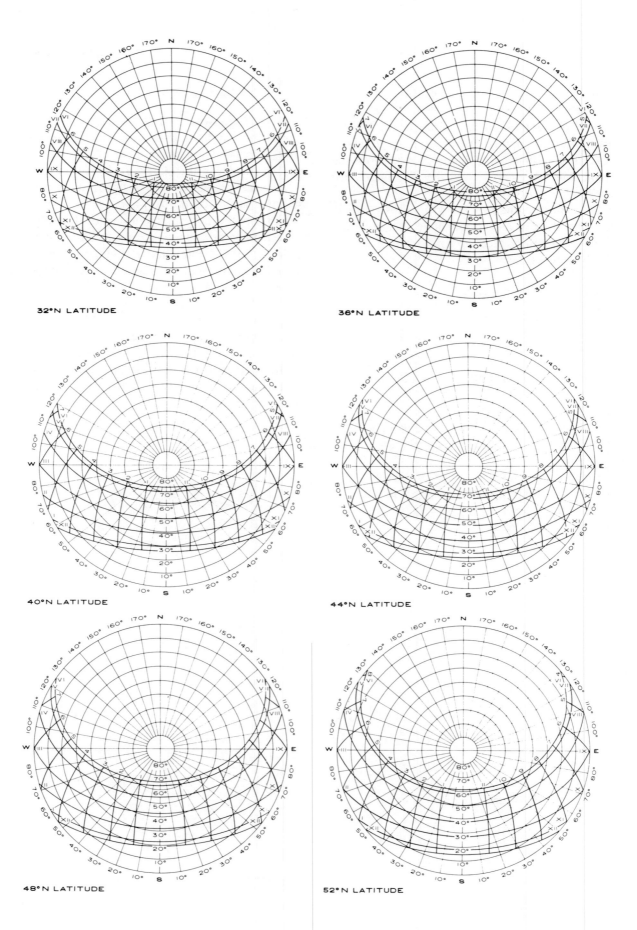

32°N LATITUDE

36°N LATITUDE

40°N LATITUDE

44°N LATITUDE

48°N LATITUDE

52°N LATITUDE

The shading mask protractor shown here can be used to construct shading masks characteristic of various shading devices. The bottom half of the protractor is used for constructing the segmental shading masks characteristic of horizontal devices (such as overhangs), as explained earlier. The upper half, turned around so the the 0° arrow points down (south) is used to construct the radial shading masks characteristic of vertical devices. These masks can be superimposed on the appropriate sun path diagram to determine the times when a surface will be shaded by these shading devices. (Source: Ramsey and Sleeper, *Architectural Graphic Standards.* Wiley.)

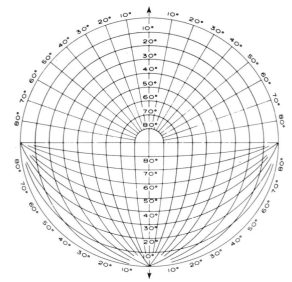

SHADING MASK PROTRACTOR

2

Clear Day Insolation Data

ASHRAE has developed tables that give the clear day insolation on tilted and south facing surfaces, such as those commonly used for solar collectors. For north latitudes (L) equal to 24°, 32°, 40°, 48°, and 56°, insolation values are given for south facing surfaces with tilt angles equal to L − 10°, L, L + 10° L + 20°, and 90° (vertical). Values are also given for the direct normal (perpendicular to the sun's rays) radiation and the insolation on a horizontal surface. The values listed in these tables are the sum of the direct solar and diffuse sky radiation hitting each surface on an average cloudless day. Data are given for the 21st day of each month; both hourly and daily total insolation are provided.

A brief examination of the 24° N table reveals that the insolation of south-facing surfaces is symmetrical about solar noon. The values given for 8 a.m. are the same as those for 4 p.m., and they are listed concurrently. Moving from left to right on any fixed time line, you encounter values of: the solar altitude and azimuth in degrees: the direct normal radiation and the insolation on a horizontal surface in Btu/(hr ft²); and the insolation of the five south facing surfaces discussed above in Btu/(hr ft²). Below these hourly data are values of the daily total insolation for each of these surfaces (in Btu/ft²). An example will help to illustrate the use of these tables.

*Example:*Determine the optimum tilt angle for a flat plate collector located in Atlanta, Georgia (32° N latitude). Select the tilt angle to maximize the surface insolation for the following three periods: a) heating season, b) cooling season, and c) the full year.

1) The heating season in Atlanta lasts from October through April: the cooling season from May to September.

2) Using the 32° N table, we sum the surface daily totals for the 22° tilt for the months October through April, and get 14,469 Btu/ft². We do the same for the 32°, 42°, 52°, and 90° tilts and get totals of 15,142; 15,382; 15,172; and 10,588.

3) Comparing these totals, we conclude that the 42° tilt, or latitude + 10°, is the best orientation for solar collection during the heating season.

4) A similar set of totals is generated for the cooling season, using the data for the months May through September. These are 11,987 Btu/ft² for 22°; 11,372 for 32°; 10,492 for 42°; 9,320 for 52°; and 3,260 for 90° tilt.

5) Comparing these totals, we conclude that the 22° tilt, or latitude − 10°, is the best for summer cooling.

6) Using the data for the whole year, we get totals of: 24,456 Btu/ft² for 22°; 26,514 for 32°; 25,874 for 42°; 24,492 for 52°; and 13,848 for 90° tilt.

7) Comparing these totals, we choose the 32° tilt, or latitude, as the best for year-round collection.

These conclusions are useful for the designer as they stand, but a little closer scrutiny is instructive. For example, the 42° tilt is best for heating, but the heating season totals for 32° and 52° are within 2 percent of the 42° total. Thus, other design considerations (such as building layout, structural framing, height restrictions) can enter the decision process without seriously affecting the final collector efficiency.

The Clear Day Insolation Data are an extremely valuable design tool, but their limitations should be kept in mind. For instance, there is no ground reflection included in the listed values. This can lead one to underestimate the clear day insolation on a vertical surface. In the example above, the heating season total for a 90° surface is about 30 percent below the 42° maximum. In reality, the insolation on a vertical surface is only 10 to 20 percent lower than this maximum during the heating season because of the contribution of these data is their assumption of an "average" clear day. Many locations are clearer than this (high altitudes and deserts), and many are less clear (industrial and dusty areas). To correct for this assumption, the numbers in these tables should be multiplied by the area-wide clearness factors listed in the ASHRAE *Handbook of Fundamentals*. Finally, the Clear Day Insolation Data do not account for cloudy weather conditions, which become quite important for long term predictions. (Source: Morrison and Farber, "Development and Use of Solar Insolation Data in Northern Latitudes for South Facing Surfaces," symposium paper in *Solar Energy Applications*, ASHRAE. Used by permission.)

SOLAR POSITION AND INSOLATION, 24°N LATITUDE

DATE	AM	PM	SOLAR POSITION ALT	SOLAR POSITION AZM	NORMAL	HORIZ.	14	24	34	54	90
JAN 21	7	5	4.8	65.6	71	10	17	21	25	28	31
	8	4	16.9	58.3	239	83	110	126	137	145	127
	9	3	27.9	48.8	288	151	188	207	221	228	176
	10	2	37.2	36.1	308	204	246	268	282	287	207
	11	1	43.6	19.6	317	237	283	306	319	324	226
	12		46.0	0.0	320	249	296	319	332	336	232
	SURFACE DAILY TOTALS				2766	1622	1984	2174	2300	2360	1766
FEB 21	7	5	9.3	74.6	158	35	44	49	53	56	46
	8	4	22.3	67.2	263	116	135	145	150	151	102
	9	3	34.4	57.6	298	187	213	225	230	228	141
	10	2	45.1	44.2	314	241	273	286	291	287	168
	11	1	53.0	25.0	321	276	310	324	328	323	185
	12		56.0	0.0	324	288	323	337	341	335	191
	SURFACE DAILY TOTALS				3036	1998	2276	2396	2424	2360	1476
MAR 21	7	5	13.7	83.8	194	60	63	64	62	59	27
	8	4	27.2	76.8	267	141	150	152	149	142	64
	9	3	40.2	67.9	295	212	226	229	225	214	95
	10	2	52.3	54.8	309	266	285	288	283	270	120
	11	1	61.9	33.4	315	300	322	326	320	305	135
	12		66.0	0.0	317	312	334	339	333	317	140
	SURFACE DAILY TOTALS				3078	2270	2428	2456	2412	2298	1022
APR 21	6	6	4.7	100.6	40	7	5	4	4	3	2
	7	5	18.3	94.9	203	83	77	70	62	51	10
	8	4	32.0	89.0	256	160	157	149	137	122	16
	9	3	45.6	81.9	280	227	227	220	206	186	41
	10	2	59.0	71.8	292	278	282	275	259	237	61
	11	1	71.1	51.6	298	310	316	309	293	269	74
	12		77.6	0.0	299	321	328	321	305	280	79
	SURFACE DAILY TOTALS				3036	2454	2458	2374	2228	2016	488
MAY 21	6	6	8.0	108.4	86	22	15	10	9	9	5
	7	5	21.2	103.2	203	98	85	73	59	44	12
	8	4	34.6	98.5	248	171	159	145	127	106	15
	9	3	48.3	93.6	269	233	224	210	190	165	16
	10	2	62.0	87.7	280	281	275	261	239	211	22
	11	1	75.5	76.9	286	311	307	293	270	240	34
	12		86.0	0.0	288	322	317	304	281	250	37
	SURFACE DAILY TOTALS				3032	2556	2447	2286	2072	1800	246
JUN 21	6	6	9.3	111.6	97	29	20	12	11	9	7
	7	5	22.3	106.8	201	103	87	73	58	41	13
	8	4	35.5	102.6	242	173	158	142	122	99	16
	9	3	49.0	98.7	263	234	221	204	182	155	18
	10	2	62.6	95.0	274	280	269	253	229	199	18
	11	1	76.3	90.8	279	309	300	283	259	227	19
	12		89.4	0.0	281	319	310	294	269	236	22
	SURFACE DAILY TOTALS				2994	2574	2422	2250	1992	1700	204

DATE	AM	PM	SOLAR POSITION ALT	SOLAR POSITION AZM	NORMAL	HORIZ.	14	24	34	54	90
JUL 21	6	6	8.2	109.0	81	23	16	11	10	9	6
	7	5	21.4	103.8	195	98	85	73	59	44	13
	8	4	34.8	99.2	239	169	157	143	125	104	16
	9	3	48.4	94.5	261	231	221	207	187	161	18
	10	2	62.1	89.0	272	278	270	256	235	206	21
	11	1	75.7	79.2	278	307	302	287	265	235	32
	12		86.6	0.0	280	317	312	298	275	245	36
	SURFACE DAILY TOTALS				2932	2526	2412	2250	2036	1766	246
AUG 21	6	6	5.0	101.3	35	7	5	4	4	4	2
	7	5	18.5	95.6	186	82	76	69	60	50	11
	8	4	32.2	89.7	241	158	154	146	134	118	16
	9	3	45.9	82.9	265	223	222	214	200	181	39
	10	2	59.3	73.0	278	273	275	268	252	230	58
	11	1	71.6	53.2	284	304	309	301	285	261	71
	12		78.3	0.0	286	315	320	313	296	272	75
	SURFACE DAILY TOTALS				2864	2408	2402	2316	2168	1958	470
SEP 21	7	5	13.7	83.8	173	57	60	60	59	56	26
	8	4	27.2	76.8	248	136	144	146	143	136	62
	9	3	40.2	67.9	278	205	218	221	217	206	93
	10	2	52.3	54.8	292	258	275	278	273	261	116
	11	1	61.9	33.4	299	291	311	315	309	295	131
	12		66.0	0.0	301	302	323	327	321	306	136
	SURFACE DAILY TOTALS				2878	2194	2342	2366	2322	2212	992
OCT 21	7	5	9.1	74.1	138	32	40	45	48	50	42
	8	4	22.0	66.7	247	111	129	139	144	145	99
	9	3	34.1	57.1	284	180	206	217	223	221	138
	10	2	44.7	43.8	301	234	265	277	282	279	165
	11	1	52.5	24.7	309	268	301	315	319	314	182
	12		55.5	0.0	311	279	314	328	332	327	188
	SURFACE DAILY TOTALS				2868	1928	2198	2314	2346	2364	1442
NOV 21	7	5	4.9	65.8	67	10	16	20	24	27	29
	8	4	17.0	58.4	232	82	108	123	135	142	124
	9	3	28.0	48.9	282	150	186	205	217	224	172
	10	2	37.3	36.3	303	203	244	265	278	283	204
	11	1	43.8	19.7	312	236	280	302	316	320	222
	12		46.2	0.0	315	247	293	315	328	332	228
	SURFACE DAILY TOTALS				2706	1610	1962	2146	2268	2324	1730
DEC 21	7	5	3.2	62.6	30	7	9	11	12	12	14
	8	4	14.9	55.3	225	71	99	116	129	139	130
	9	3	25.5	46.0	281	137	176	198	214	223	184
	10	2	34.3	33.7	304	189	234	258	275	283	217
	11	1	40.4	18.2	314	221	270	295	312	320	236
	12		42.6	0.0	317	232	282	308	325	332	243
	SURFACE DAILY TOTALS				2624	1474	1852	2058	2204	2286	1808

Column notes: NORMAL and HORIZ. are BTUH/SQ. FT. TOTAL INSOLATION ON SURFACES. Columns 14, 24, 34, 54, 90 = SOUTH FACING SURFACE ANGLE WITH HORIZ.

SOLAR POSITION AND INSOLATION, 32°N LATITUDE

JAN 21 – JUN 21

DATE	SOLAR TIME AM	SOLAR TIME PM	ALT	AZM	NORMAL	HORIZ.	22	32	42	52	90
JAN 21	7	5	1.4	65.2	203	56	93	106	116	123	115
	8	4	12.5	56.5	269	118	175	193	206	212	181
	9	3	22.5	46.0	295	167	235	256	269	274	221
	10	2	30.6	33.1	306	198	273	295	308	312	245
	11	1	36.1	17.5	310	209	285	308	321	324	253
	12		38.0	0.0						1	1
	SURFACE DAILY TOTALS				2958	1288	1839	2008	2118	2166	1779
FEB 21	7	5	7.1	73.5	121	22	34	37	40	42	38
	8	4	19.0	64.4	247	95	127	136	140	141	108
	9	3	29.9	53.4	288	161	206	217	222	220	158
	10	2	39.1	39.4	306	212	266	278	283	279	193
	11	1	45.6	21.4	315	244	304	317	321	315	214
	12		48.0	0.0	317	255	316	330	334	328	222
	SURFACE DAILY TOTALS				2882	1724	2188	2300	2345	2322	1644
MAR 21	7	5	12.7	81.9	185	54	60	60	59	56	32
	8	4	25.1	73.0	260	129	146	147	144	137	78
	9	3	36.8	62.1	290	194	222	224	220	209	119
	10	2	47.3	47.5	304	245	280	283	278	265	150
	11	1	55.0	26.8	311	277	317	321	315	300	170
	12		58.0	0.0	313	287	329	333	327	312	177
	SURFACE DAILY TOTALS				3012	2084	2378	2403	2358	2246	1276
APR 21	6	6	6.1	99.9	66	14	9	6	6	5	3
	7	5	18.8	92.2	206	86	78	71	62	51	10
	8	4	31.5	84.0	255	158	156	148	136	120	35
	9	3	43.9	74.2	278	220	225	217	203	183	68
	10	2	55.7	60.3	290	267	279	272	256	234	95
	11	1	65.4	37.5	295	297	313	306	290	265	112
	12		69.6	0.0	297	307	325	318	301	276	118
	SURFACE DAILY TOTALS				3076	2390	2444	2356	2206	1994	764
MAY 21	6	6	10.4	107.2	119	36	21	16	13	12	13
	7	5	22.8	100.1	211	107	88	75	60	44	15
	8	4	35.4	92.9	250	175	159	145	127	105	33
	9	3	48.1	84.7	269	233	223	209	188	163	56
	10	2	60.6	73.3	280	277	273	259	237	208	72
	11	1	72.0	51.9	285	305	305	290	268	237	77
	12		78.0	0.0	286	315	315	301	278	247	77
	SURFACE DAILY TOTALS				3112	2582	2454	2284	2064	1788	469
JUN 21	6	6	12.2	110.2	131	45	26	16	15	14	9
	7	5	24.3	103.4	210	115	91	76	59	41	14
	8	4	36.9	96.8	245	180	159	143	122	99	16
	9	3	49.6	89.4	264	236	221	204	181	153	19
	10	2	62.2	79.7	274	279	268	251	227	197	41
	11	1	74.2	60.9	279	306	299	282	257	224	56
	12		81.5	0.0	280	315	309	292	267	234	60
	SURFACE DAILY TOTALS				3084	2634	2436	2234	1990	1690	370

JUL 21 – DEC 21

DATE	SOLAR TIME AM	SOLAR TIME PM	ALT	AZM	NORMAL	HORIZ.	22	32	42	52	90
JUL 21	6	6	10.7	107.7	113	37	22	14	13	12	8
	7	5	23.1	100.6	203	107	87	75	60	44	14
	8	4	35.7	93.6	241	174	158	143	125	104	16
	9	3	48.4	85.5	261	231	220	205	185	159	31
	10	2	60.9	74.3	271	274	269	254	232	204	54
	11	1	72.4	53.3	277	302	300	285	262	232	69
	12		78.6	0.0	279	311	310	296	273	242	74
	SURFACE DAILY TOTALS				3012	2558	2422	2250	2030	1754	458
AUG 21	6	6	6.5	100.5		14	9	7	6	6	4
	7	5	19.1	92.8	190	85	77	69	60	50	12
	8	4	31.8	84.7	240	156	152	144	132	116	33
	9	3	44.3	75.0	263	216	220	212	197	178	65
	10	2	56.1	61.3	276	262	272	264	249	226	91
	11	1	66.0	38.4	282	292	305	298	281	257	107
	12		70.3	0.0	284	302	317	309	292	268	113
	SURFACE DAILY TOTALS				2902	2352	2388	2296	2144	1934	736
SEP 21	7	5	12.7	81.9	163	51	56	56	55	52	30
	8	4	25.1	73.0	240	124	140	141	138	131	75
	9	3	36.8	62.1	272	188	213	215	211	201	114
	10	2	47.3	47.5	287	237	270	273	268	255	145
	11	1	55.0	26.8	294	268	306	309	303	289	164
	12		58.0	0.0	296	278	318	321	315	300	171
	SURFACE DAILY TOTALS				2808	2014	2288	2308	2264	2154	1226
OCT 21	7	5	6.8	73.1	99	19	29	32	34	36	32
	8	4	18.7	64.0	229	90	120	128	133	134	104
	9	3	29.5	53.0	273	155	198	208	213	212	153
	10	2	38.7	39.1	293	204	257	269	273	270	188
	11	1	45.1	21.1	302	236	294	307	311	306	209
	12		47.5	0.0	304	247	306	320	324	318	217
	SURFACE DAILY TOTALS				2696	1654	2100	2208	2252	2232	1588
NOV 21	7	5	1.5	65.4	2	0	0	0	1	1	1
	8	4	12.7	56.6	196	55	91	104	113	119	111
	9	3	22.6	46.1	263	118	173	190	202	208	176
	10	2	30.8	33.2	289	166	233	252	265	270	217
	11	1	36.2	17.6	301	197	270	291	303	307	241
	12		38.2	0.0	304	207	282	304	316	320	249
	SURFACE DAILY TOTALS				2406	1280	1816	1980	2084	2130	1742
DEC 21	8	4	10.3	53.8	176	41	77	90	101	108	107
	9	3	19.8	43.6	257	102	161	180	195	204	183
	10	2	27.6	31.2	288	150	221	244	259	267	226
	11	1	32.7	16.4	301	180	258	282	298	305	251
	12		34.6	0.0	304	190	271	295	311	318	259
	SURFACE DAILY TOTALS				2348	1136	1704	1888	2016	2086	1794

BTUH/SQ. FT. — TOTAL INSOLATION ON SURFACES — SOUTH FACING SURFACE ANGLE WITH HORIZ.

SOLAR POSITION AND INSOLATION, 40°N LATITUDE

DATE	AM	PM	ALT	AZM	NORMAL	HORIZ.	30	40	50	60	90
			SOLAR POSITION		BTUH/SQ. FT. TOTAL INSOLATION ON SURFACES		SOUTH FACING SURFACE ANGLE WITH HORIZ.				
JAN 21	8	4	8.1	55.3	142	28	65	74	81	85	84
	9	3	16.8	44.0	239	83	155	171	182	187	171
	10	2	23.8	30.9	274	127	218	237	249	254	223
	11	1	28.4	16.0	289	154	257	277	290	293	253
	12		30.0	0.0	294	164	270	291	303	306	263
	SURFACE DAILY TOTALS				2182	948	1660	1810	1906	1944	1726
FEB 21	7	5	4.8	72.7	69	10	19	21	23	24	22
	8	4	15.4	62.2	224	73	114	122	126	127	107
	9	3	25.0	50.2	274	132	195	205	209	208	167
	10	2	32.8	35.9	295	178	256	267	271	267	210
	11	1	38.1	18.9	305	206	293	306	310	304	236
	12		40.0	0.0	308	216	306	319	323	317	245
	SURFACE DAILY TOTALS				2640	1414	2060	2162	2202	2176	1730
MAR 21	7	5	11.4	80.2	171	46	55	55	54	51	35
	8	4	22.5	69.6	250	114	140	141	138	131	89
	9	3	32.8	57.3	282	173	215	217	213	202	138
	10	2	41.6	41.9	297	218	273	276	271	258	176
	11	1	47.7	22.6	305	247	310	313	307	293	200
	12		50.0	0.0	307	257	322	326	320	305	208
	SURFACE DAILY TOTALS				2916	1852	2308	2330	2284	2174	1484
APR 21	6	6	7.4	98.9	89	20	11	8	7	7	4
	7	5	18.9	89.5	206	87	77	70	61	50	12
	8	4	30.3	79.3	252	152	153	145	133	117	53
	9	3	41.3	67.2	274	207	221	213	199	179	93
	10	2	51.2	51.4	286	250	275	267	252	229	126
	11	1	58.7	29.2	292	277	308	301	285	260	147
	12		61.6	0.0	293	287	320	313	296	271	154
	SURFACE DAILY TOTALS				3092	2274	2412	2320	2158	1956	1022
MAY 21	5	7	1.9	114.7	1	0	0	0	0	0	0
	6	6	12.7	105.6	144	49	25	15	14	13	9
	7	5	24.0	96.6	216	114	89	76	60	44	13
	8	4	35.4	87.2	250	175	158	144	125	104	25
	9	3	46.8	76.0	267	227	221	206	186	160	60
	10	2	57.5	60.9	277	267	270	255	233	205	89
	11	1	66.2	37.1	283	293	301	287	264	234	108
	12		70.0	0.0	284	301	312	297	274	243	114
	SURFACE DAILY TOTALS				3160	2552	2442	2264	2040	1760	724
JUN 21	5	7	4.2	117.3	22	4	3	3	2	2	1
	6	6	14.8	108.4	155	60	30	18	17	16	10
	7	5	26.0	99.7	216	123	92	77	59	41	14
	8	4	37.4	90.7	246	182	159	142	121	97	16
	9	3	48.8	80.2	263	233	219	202	179	151	47
	10	2	59.8	65.8	272	272	266	248	224	194	74
	11	1	69.2	41.9	277	296	296	278	253	221	92
	12		73.5	0.0	279	304	306	289	263	230	98
	SURFACE DAILY TOTALS				3180	2648	2434	2224	1974	1670	610

DATE	AM	PM	ALT	AZM	NORMAL	HORIZ.	30	40	50	60	90
			SOLAR POSITION		BTUH/SQ. FT. TOTAL INSOLATION ON SURFACES		SOUTH FACING SURFACE ANGLE WITH HORIZ.				
JUL 21	5	7	2.3	115.2	2	0	0	0	0	0	0
	6	6	13.1	106.1	138	50	26	17	15	14	9
	7	5	24.3	97.2	208	114	89	75	60	44	14
	8	4	35.8	87.8	241	174	157	142	124	102	24
	9	3	47.2	76.7	259	225	218	203	182	157	58
	10	2	57.9	61.7	269	265	266	251	229	200	86
	11	1	66.7	37.9	275	290	296	281	258	228	104
	12		70.6	0.0	276	298	307	292	269	238	111
	SURFACE DAILY TOTALS				3062	2534	2409	2230	2006	1728	702
AUG 21	6	6	7.9	99.5	81	21	12	9	8	7	5
	7	5	19.3	90.0	191	87	76	69	60	49	12
	8	4	30.7	79.9	237	150	150	141	129	113	50
	9	3	41.8	67.9	260	205	216	207	193	173	89
	10	2	51.7	52.1	272	246	267	259	244	221	120
	11	1	59.3	29.7	278	273	300	292	276	252	140
	12		62.3	0.0	280	282	311	303	287	262	147
	SURFACE DAILY TOTALS				2916	2244	2354	2258	2104	1894	978
SEP 21	7	5	11.4	80.2	149	43	51	51	49	47	32
	8	4	22.5	69.6	230	109	133	134	131	124	84
	9	3	32.8	57.3	263	167	206	208	203	193	132
	10	2	41.6	41.9	280	211	262	265	260	247	168
	11	1	47.7	22.6	287	239	298	301	295	281	192
	12		50.0	0.0	290	249	310	313	307	292	200
	SURFACE DAILY TOTALS				2708	1788	2210	2228	2182	2074	1416
OCT 21	7	5	4.5	72.3	48	7	14	15	17	17	16
	8	4	15.0	61.9	204	68	106	113	117	118	100
	9	3	24.5	49.8	257	126	185	195	200	198	160
	10	2	32.4	35.6	280	170	245	257	261	257	203
	11	1	37.6	18.7	291	199	283	295	299	294	229
	12		39.5	0.0	294	208	295	308	312	306	238
	SURFACE DAILY TOTALS				2454	1348	1962	2060	2098	2074	1654
NOV 21	8	4	8.2	55.4	136	28	63	72	78	82	81
	9	3	17.0	44.1	232	82	152	167	178	183	167
	10	2	24.0	31.0	268	126	215	233	245	249	219
	11	1	28.6	16.1	283	153	254	273	285	288	248
	12		30.2	0.0	288	163	267	287	298	301	258
	SURFACE DAILY TOTALS				2128	942	1636	1778	1870	1908	1686
DEC 21	8	4	5.5	53.0	89	14	39	45	50	54	56
	9	3	14.0	41.9	217	65	135	157	164	171	163
	10	2	20.7	29.4	261	107	200	221	235	242	221
	11	1	25.0	15.2	280	134	239	262	276	283	252
	12		26.6	0.0	285	143	253	275	290	296	263
	SURFACE DAILY TOTALS				1978	782	1480	1634	1740	1796	1646

SOLAR POSITION AND INSOLATION, 48°N LATITUDE

DATE	AM	PM	SOLAR POSITION ALT	SOLAR POSITION AZM	NORMAL	HORIZ.	38	48	58	68	90
							SOUTH FACING SURFACE ANGLE WITH HORIZ.				
JAN 21	8	4	3.5	54.6	37	4	17	19	21	22	22
	9	3	11.0	42.6	185	46	120	132	140	145	139
	10	2	16.9	29.4	239	83	190	206	216	220	206
	11	1	20.7	15.1	261	107	231	249	260	263	243
	12		22.0	0.0	267	115	245	264	275	278	255
	SURFACE DAILY TOTALS				1710	596	1360	1478	1550	1578	1478
FEB 21	7	5	2.4	72.2	12	1	3	4	4	4	1
	8	4	11.6	60.5	188	49	95	102	105	106	96
	9	3	19.7	47.7	251	100	178	187	191	190	167
	10	2	26.2	33.3	278	139	240	251	255	251	217
	11	1	30.5	17.2	290	165	278	290	294	288	247
	12		32.0	0.0	293	173	291	304	307	301	258
	SURFACE DAILY TOTALS				2330	1080	1880	1972	2024	1978	1720
MAR 21	7	5	10.0	78.7	153	37	49	49	47	45	35
	8	4	19.5	66.8	236	96	131	132	129	122	96
	9	3	28.2	53.4	270	147	205	207	203	193	152
	10	2	35.4	37.8	287	187	263	266	261	248	195
	11	1	40.3	19.8	295	212	300	303	297	283	223
	12		42.0	0.0	298	220	312	315	309	294	232
	SURFACE DAILY TOTALS				2780	1578	2208	2228	2182	2074	1632
APR 21	6	6	8.6	97.8	108	27	13	9	8	7	5
	7	5	18.6	86.7	205	85	76	69	59	48	21
	8	4	28.5	74.9	247	142	149	141	129	113	69
	9	3	37.8	61.2	268	191	216	208	194	174	115
	10	2	45.8	44.6	280	228	268	260	245	223	152
	11	1	51.5	24.0	286	252	301	294	278	254	177
	12		53.6	0.0	288	260	313	305	289	264	185
	SURFACE DAILY TOTALS				3076	2106	2358	2266	2114	1902	1262
MAY 21	5	7	5.2	114.3	41	9	4	3	4	3	2
	6	6	14.7	103.7	162	61	27	16	15	13	10
	7	5	24.6	93.0	219	118	89	75	60	43	13
	8	4	34.7	81.6	248	171	156	142	123	101	45
	9	3	44.3	68.3	264	217	217	202	182	156	86
	10	2	53.0	51.3	274	252	265	251	229	200	120
	11	1	59.5	28.6	279	274	296	281	258	228	141
	12		62.0	0.0	280	281	306	292	269	238	149
	SURFACE DAILY TOTALS				3254	2482	2418	2234	2010	1728	982
JUN 21	5	7	7.9	116.5	77	21	9	7	8	7	5
	6	6	17.2	106.2	172	74	33	19	18	16	12
	7	5	27.0	95.8	220	129	93	77	59	39	15
	8	4	37.1	84.6	246	181	157	140	119	95	35
	9	3	46.9	71.6	261	225	216	198	175	147	74
	10	2	55.8	54.8	269	259	262	244	221	189	105
	11	1	62.7	31.2	274	280	291	273	248	216	126
	12		65.5	0.0	275	287	301	283	258	225	133
	SURFACE DAILY TOTALS				3312	2626	2420	2204	1950	1644	874

DATE	AM	PM	SOLAR POSITION ALT	SOLAR POSITION AZM	NORMAL	HORIZ.	38	48	58	68	90
							SOUTH FACING SURFACE ANGLE WITH HORIZ.				
JUL 21	5	7	5.7	114.7	43	10	5	5	4	4	3
	6	6	15.2	104.1	156	62	28	18	16	15	11
	7	5	25.1	93.5	211	118	89	75	59	42	14
	8	4	35.1	82.1	240	171	154	140	121	99	43
	9	3	44.8	68.8	256	215	214	199	178	153	83
	10	2	53.5	51.9	266	250	261	246	224	195	116
	11	1	60.1	29.0	271	272	291	276	253	223	137
	12		62.6	0.0	272	279	301	286	263	232	144
	SURFACE DAILY TOTALS				3158	2474	2386	2200	1974	1694	956
AUG 21	6	6	9.1	98.3	99	28	14	10	9	8	6
	7	5	19.1	87.2	190	85	75	67	58	47	20
	8	4	29.0	75.4	232	141	145	137	125	109	65
	9	3	38.4	61.8	254	189	210	201	187	168	110
	10	2	46.4	45.1	266	225	260	252	237	214	146
	11	1	52.4	24.3	272	248	293	285	268	244	169
	12		54.3	0.0	274	256	304	296	279	255	177
	SURFACE DAILY TOTALS				2898	2086	2300	2200	2046	1836	1208
SEP 21	7	5	10.0	78.7	131	35	44	44	43	40	31
	8	4	19.5	66.8	215	92	124	124	121	115	90
	9	3	28.2	53.4	251	142	196	197	193	183	143
	10	2	35.4	37.8	269	181	251	254	248	236	185
	11	1	40.3	19.8	278	205	287	289	284	269	212
	12		42.0	0.0	280	213	299	302	296	281	221
	SURFACE DAILY TOTALS				2568	1522	2102	2118	2070	1966	1546
OCT 21	7	5	2.0	71.9	4	0	4	1	1	1	1
	8	4	11.2	60.2	165	44	86	91	95	95	87
	9	3	19.3	47.4	233	94	167	176	180	178	157
	10	2	25.7	33.1	262	133	228	239	242	239	207
	11	1	30.0	17.1	274	157	266	277	281	276	237
	12		31.5	0.0	278	166	279	291	294	288	247
	SURFACE DAILY TOTALS				2154	1022	1774	1860	1890	1866	1626
NOV 21	8	4	3.6	54.7	36	5	17	19	21	22	22
	9	3	11.2	42.7	179	46	117	129	137	141	135
	10	2	17.1	29.5	233	83	186	202	212	215	201
	11	1	20.9	15.1	255	107	227	245	255	258	238
	12		22.2	0.0	261	115	241	259	270	272	250
	SURFACE DAILY TOTALS				1668	596	1336	1448	1518	1544	1442
DEC 21	9	3	8.0	40.9	140	27	87	98	105	110	109
	10	2	13.6	28.2	214	63	164	180	192	197	190
	11	1	17.3	14.4	242	86	207	226	239	244	231
	12		18.6	0.0	250	94	222	241	254	260	244
	SURFACE DAILY TOTALS				1442	446	1136	1250	1326	1364	1304

BTUH/SQ. FT. TOTAL INSOLATION ON SURFACES

SOLAR POSITION AND INSOLATION, 56°N LATITUDE

DATE	AM	PM	ALT	AZM	NORMAL	HORIZ.	46	56	66	76	90
JAN 21	9	3	5.0	41.8	78	11	50	55	59	60	60
	10	2	9.9	28.5	170	39	135	146	154	156	153
	11	1	12.9	14.5	207	58	183	197	206	208	201
	12		14.0	0.0	217	65	198	214	222	225	217
	SURFACE DAILY TOTALS				1126	282	934	1010	1058	1074	1044
FEB 21	8	4	7.6	59.4	129	25	65	69	72	72	69
	9	3	14.2	45.9	214	65	151	159	162	161	151
	10	2	19.4	31.5	250	98	215	225	228	224	208
	11	1	22.8	16.1	266	119	254	265	268	263	243
	12		24.0	0.0	270	126	268	279	282	276	255
	SURFACE DAILY TOTALS				1986	740	1640	1716	1742	1716	1598
MAR 21	7	5	8.3	77.5	128	28	40	40	39	37	32
	8	4	16.2	64.4	215	75	119	120	117	111	97
	9	3	23.3	50.3	253	118	192	193	189	180	154
	10	2	29.0	34.9	272	151	249	251	246	234	205
	11	1	32.7	17.9	282	172	285	288	282	268	236
	12		34.0	0.0	284	179	297	300	294	280	246
	SURFACE DAILY TOTALS				2586	1268	2066	2084	2040	1938	1700
APR 21	5	7	1.4	108.8	0	0	0	0	0	0	0
	6	6	9.6	96.5	122	32	14	9	8	7	6
	7	5	18.0	84.1	201	81	74	66	57	46	29
	8	4	26.1	70.9	239	129	143	135	123	108	82
	9	3	33.6	56.3	260	169	208	200	186	167	133
	10	2	39.9	39.7	272	201	259	251	236	214	174
	11	1	44.1	20.7	278	220	292	285	268	245	200
	12		45.6	0.0	280	227	303	295	279	255	209
	SURFACE DAILY TOTALS				3024	1892	2282	2186	2038	1830	1458
MAY 21	4	8	1.2	125.5	0	0	0	0	0	0	0
	5	7	8.5	113.4	93	25	10	9	8	7	6
	6	6	16.5	101.5	175	71	28	17	15	13	11
	7	5	24.8	89.3	219	119	88	74	58	41	16
	8	4	33.1	76.3	244	163	153	138	119	98	63
	9	3	40.9	61.6	259	201	212	197	176	151	109
	10	2	47.6	44.2	268	231	259	244	222	194	146
	11	1	52.3	23.4	273	249	288	274	251	222	170
	12		54.0	0.0	275	255	299	284	261	231	178
	SURFACE DAILY TOTALS				3340	2374	2374	2188	1962	1682	1218
JUN 21	4	8	4.2	127.2	21	4	2	2	2	2	1
	5	7	11.4	115.3	122	40	14	13	11	10	8
	6	6	19.3	103.6	185	86	34	19	17	15	12
	7	5	27.6	91.7	222	132	92	76	57	38	15
	8	4	35.9	78.8	243	175	154	137	116	92	55
	9	3	43.8	64.1	257	212	211	193	170	143	98
	10	2	50.7	46.4	265	240	255	238	214	184	133
	11	1	55.6	24.9	269	258	284	267	242	210	156
	12		57.5	0.0	271	264	294	276	251	219	164
	SURFACE DAILY TOTALS				3458	2562	2388	2166	1910	1606	1120

DATE	AM	PM	ALT	AZM	NORMAL	HORIZ.	46	56	66	76	90
JUL 21	4	8	1.7	125.8	0	0	0	0	0	0	0
	5	7	9.0	113.7	91	27	11	10	9	8	6
	6	6	17.0	101.9	169	72	30	18	16	14	12
	7	5	25.3	89.7	212	119	88	74	58	41	15
	8	4	33.6	76.7	237	163	151	136	117	96	61
	9	3	41.4	62.0	252	201	208	193	173	147	106
	10	2	48.2	44.6	261	230	254	239	217	189	142
	11	1	52.9	23.7	265	248	283	268	245	216	165
	12		54.6	0.0	267	254	293	278	255	225	173
	SURFACE DAILY TOTALS				3240	2372	2342	2152	1926	1646	1186
AUG 21	5	7	2.0	109.2	1	0	0	0	0	0	0
	6	6	10.2	97.0	112	34	16	11	10	9	7
	7	5	18.5	84.5	187	82	73	65	56	45	28
	8	4	26.7	71.3	225	128	140	131	119	104	78
	9	3	34.3	56.7	246	168	202	193	179	160	126
	10	2	40.5	40.0	258	199	251	242	227	206	166
	11	1	44.8	20.9	264	218	282	274	258	235	191
	12		46.3	0.0	266	225	293	285	269	245	200
	SURFACE DAILY TOTALS				2850	1884	2218	2118	1966	1760	1392
SEP 21	7	5	8.3	77.5	107	25	36	36	34	32	28
	8	4	16.2	64.4	194	72	111	111	108	102	89
	9	3	23.3	50.3	233	114	181	182	178	168	147
	10	2	29.0	34.9	253	146	236	237	232	221	193
	11	1	32.7	17.9	263	166	271	273	267	254	223
	12		34.0	0.0	266	173	283	285	279	265	233
	SURFACE DAILY TOTALS				2568	1220	1950	1962	1918	1820	1594
OCT 21	8	4	7.1	59.1	104	20	53	57	59	59	57
	9	3	13.8	45.7	193	60	138	145	148	147	138
	10	2	19.0	31.3	231	92	201	210	213	210	195
	11	1	22.3	16.0	248	112	240	250	253	248	230
	12		23.5	0.0	253	119	253	263	266	261	241
	SURFACE DAILY TOTALS				1804	688	1516	1586	1612	1588	1480
NOV 21	9	3	5.2	41.9	76	12	49	54	57	59	58
	10	2	10.0	28.5	165	39	132	143	149	152	148
	11	1	13.1	14.5	201	58	179	193	201	203	196
	12		14.2	0.0	211	65	194	209	217	219	211
	SURFACE DAILY TOTALS				1094	284	914	986	1032	1046	1016
DEC 21	9	3	1.9	40.5	5	0	3	4	4	4	4
	10	2	6.6	27.5	113	19	86	95	101	104	103
	11	1	9.5	13.9	166	37	141	154	163	167	164
	12		10.6	0.0	180	43	159	173	182	186	182
	SURFACE DAILY TOTALS				748	156	620	678	716	734	722

3

Solar Radiation Maps

The quantity of solar radiation actually available for use in heating is difficult to calculate exactly. Most of this difficulty is due to the many factors that influence the radiation available at a collector location. But most of these factors can be treated by statistical methods using long-term averages of recorded weather data.

The least modified and therefore most usable solar radiation data is available from the U.S. Weather Bureau. Some of these data, averaged over a period of many years, have been published in the *Climatic Atlas of the United States* in the form of tables or maps. A selection of these average data is reprinted here for convenience. They are taken to be a good indicator of future weather trends. More recent and complete information may be obtained from the National Weather Records Center in Asheville, North Carolina.

As one example, daily insolation has been recorded at more than 80 weather stations across the United States. The available data have been averaged over a period of more than 30 years; these averages are summarized in the first 12 (one for each month) contour maps: "Mean Daily Solar Radiation." Values are given in langleys, or calories per square centimeter. Multiply by 3.69 to convert to Btu per square foot. These figures represent the monthly average of the daily total of direct, diffuse, and reflected ra-

diation on a horizontal surface. Trigonometric conversions must be applied to these data to convert them to the insolation on vertical or tilted surfaces.

Other useful information includes the Weather Bureau records of the amount of sunshine, which is listed as the "hours of sunshine" or the "percentage of possible sunshine." A device records the cumulative total hours from sunrise to sunset to get the percentage of possible sunshine. Monthly averages of this percentage are provided in the next 12 contour maps here, "Mean Percentage of Possible Sunshine." These values can be taken as the average portion of the daytime hours each month when the sun is not obscured by clouds.

Also included in each of these 12 maps is a table of the average number of hours between sunrise and sunset for that month. You can multiply this number by the mean percentage of possible sunshine to obtain the mean number of hours of sunshine for a particular month and location. A table at the end of this section lists the mean number of hours of sunshine for selected locations across the United States.

These national maps are useful for getting an overview or approximation of the available solar radiation at a particular spot. For many locations, they may be the only way of finding a particular value. As a rule, however, they should

161

be used only when other more local data are unavailable. Many local factors can have significant effect, so care and judgement are important when using interpolated data from these national weather maps. (Source: Environmental Science Services Administration, *Climatic Atlas of the United States,* U.S. Department of Commerce.)

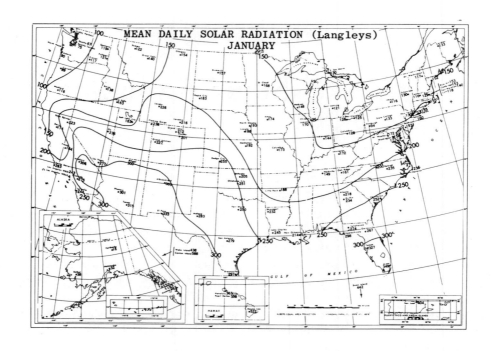

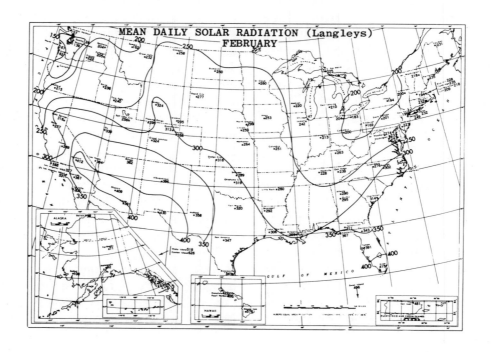

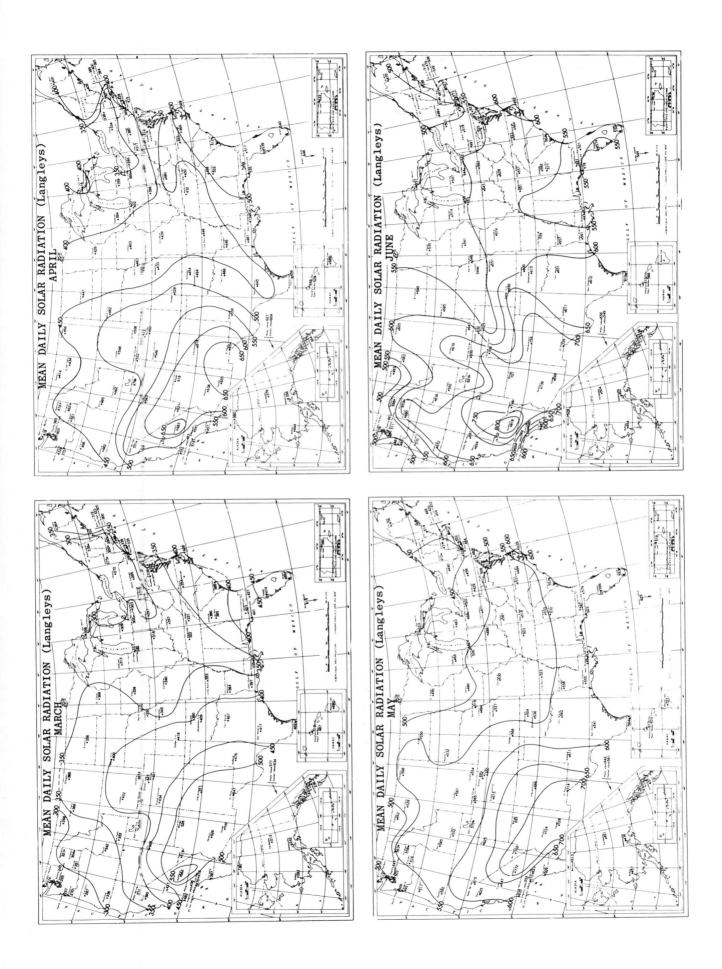

MEAN DAILY SOLAR RADIATION (Langleys)
MARCH

MEAN DAILY SOLAR RADIATION (Langleys)
APRIL

MEAN DAILY SOLAR RADIATION (Langleys)
MAY

MEAN DAILY SOLAR RADIATION (Langleys)
JUNE

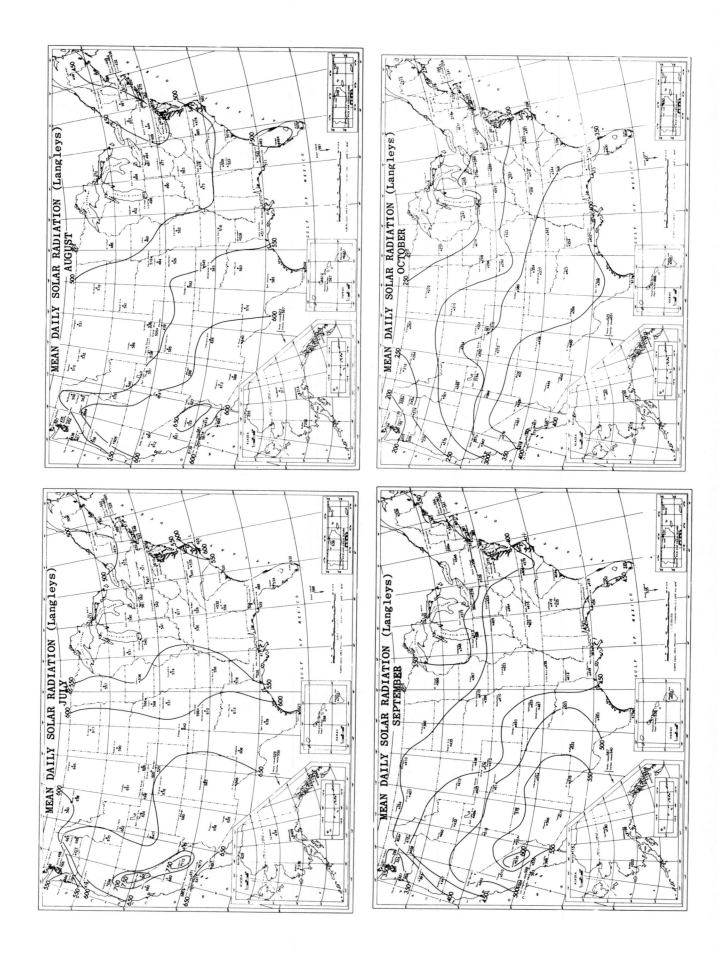

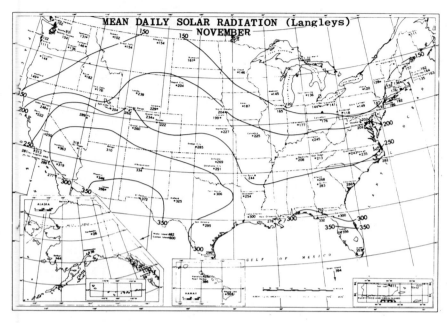

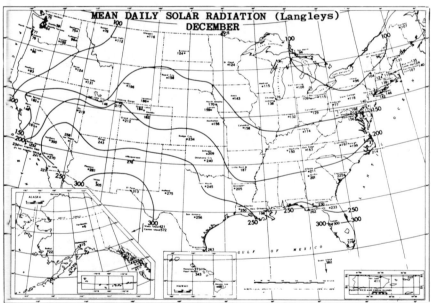

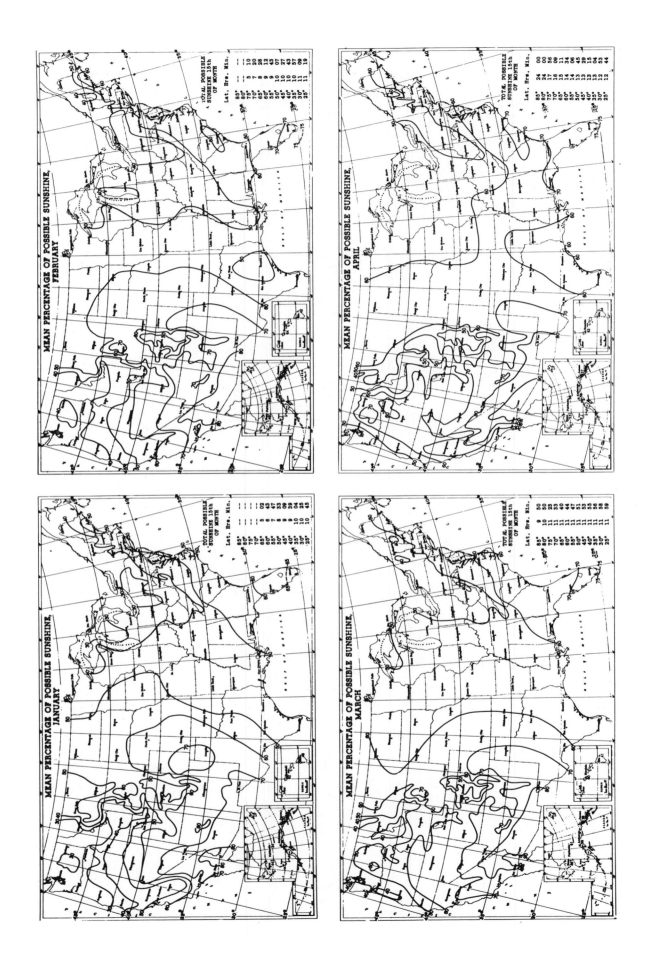

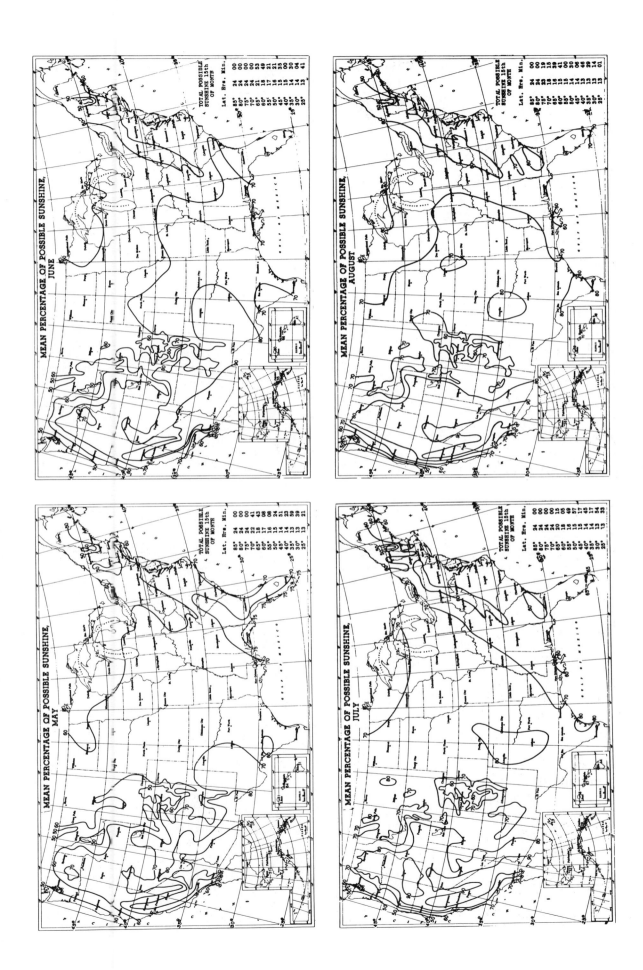

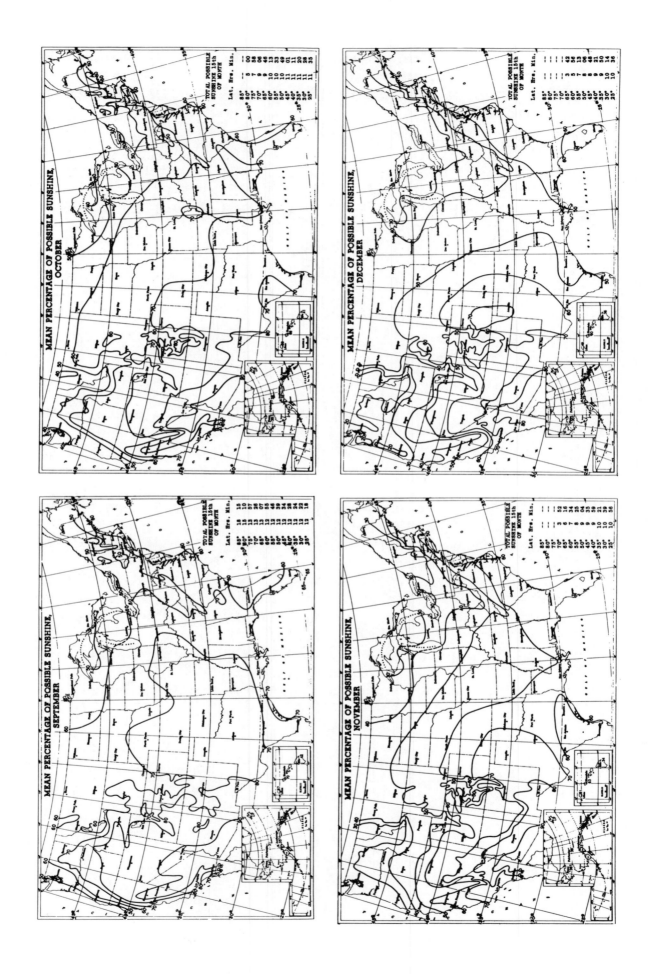

MEAN NUMBER OF HOURS OF SUNSHINE

STATE AND STATION	YEARS	JAN.	FEB.	MAR.	APR.	MAY	JUNE	JULY	AUG.	SEPT.	OCT.	NOV.	DEC.	ANNUAL
ALA. BIRMINGHAM	30	138	152	207	248	293	294	269	265	244	234	182	136	2662
MOBILE	22	157	158	212	253	301	289	249	259	235	254	195	146	2708
MONTGOMERY	30	160	168	227	267	317	311	288	290	260	250	200	156	2894
ALASKA ANCHORAGE	19	78	114	210	254	268	288	255	184	128	96	68	49	1992
FAIRBANKS	20	54	120	224	302	319	334	274	164	122	85	71	36	2105
JUNEAU	29	71	102	171	200	230	251	193	161	123	67	60	51	1680
NOME	[illegible]	78	100	190	[illegible]	285	297	204	146	142	101	67	42	1884
ARIZ. PHOENIX	30	248	244	314	346	404	404	377	351	334	307	267	236	3832
PRESCOTT	14	222	230	293	323	378	392	323	305	315	286	254	228	3549
TUCSON	13	255	266	317	350	399	394	329	329	335	317	280	258	3829
YUMA	30	258	266	337	365	419	420	404	380	351	330	285	262	4077
ARK. FT. SMITH	30	146	156	202	234	268	303	321	305	261	230	174	147	2747
LITTLE ROCK	30	143	158	213	243	291	316	321	316	265	251	181	142	2840
CALIF. EUREKA	30	120	138	180	209	247	261	244	205	195	164	127	108	2198
FRESNO	29	153	192	283	330	389	418	435	406	355	306	221	144	3632
LOS ANGELES	30	224	217	273	264	292	299	352	336	295	263	249	220	3284
RED BLUFF	15	156	186	246	302	366	396	438	407	341	277	199	154	3468
SACRAMENTO	30	134	169	255	300	367	405	437	406	347	283	197	122	3422
SAN DIEGO	30	216	212	262	242	261	253	293	277	255	234	236	217	2958
SAN FRANCISCO	30	165	182	251	281	314	330	300	272	267	243	198	156	2959
COLO. DENVER	30	207	205	247	252	281	311	321	297	274	246	200	192	3033
GRAND JUNCTION	30	169	182	243	265	314	350	349	311	291	255	198	168	3095
PUEBLO	30	224	217	261	271	299	340	349	318	290	265	225	211	3270
CONN. HARTFORD	30	141	166	206	223	267	285	299	268	220	193	137	136	2541
NEW HAVEN	30	155	178	215	234	274	291	309	284	238	215	157	154	2704
D. C. WASHINGTON	30	138	160	205	226	267	288	291	264	233	207	162	135	2576
FLA. APALACHICOLA	26	193	195	233	274	328	296	273	259	236	263	216	175	2941
JACKSONVILLE	30	192	189	241	267	296	260	255	248	199	205	191	170	2713
KEY WEST	30	229	238	285	296	307	273	277	269	236	237	226	225	3098
LAKELAND	7	204	186	222	251	285	268	252	242	203	209	212	198	2732
MIAMI	30	222	227	266	275	280	251	267	263	216	215	212	209	2903
PENSACOLA	30	175	180	232	270	311	302	278	284	249	265	206	166	2918
TAMPA	30	223	220	260	283	320	278	257	252	232	243	227	209	3001
GA. ATLANTA	25	154	165	218	266	309	304	284	285	247	241	188	160	2821
MACON	30	177	178	235	279	321	314	292	295	253	236	202	168	2950
SAVANNAH	30	175	173	229	274	307	279	267	256	212	216	197	167	2752
HAWAII HILO	7	153	135	161	112	106	158	184	134	137	153	106	131	1670
HONOLULU	30	227	202	250	255	276	280	293	290	279	257	221	211	3041
LIHUE	10	171	162	176	176	211	246	246	236	246	210	170	161	2411
IDAHO BOISE	30	116	144	218	274	322	352	412	378	311	232	143	104	3006
POCATELLO	30	111	143	211	255	300	338	380	347	296	230	145	108	2864
ILL. CAIRO	15	124	160	218	254	298	324	345	336	279	254	181	145	2918
CHICAGO	30	126	142	199	221	274	300	333	299	247	216	136	118	2611
MOLINE	18	132	139	189	214	255	279	337	300	251	214	130	123	2563
PEORIA	30	134	149	198	229	273	303	336	299	259	222	149	122	2673
SPRINGFIELD	30	127	149	193	224	282	304	346	312	266	225	152	122	2702
IND. EVANSVILLE	30	123	145	199	237	294	322	342	318	274	236	156	120	2766
FT. WAYNE	30	113	136	191	217	281	310	342	306	242	210	120	102	2570
INDIANAPOLIS	30	118	140	193	227	278	313	342	313	265	222	139	118	2668
TERRE HAUTE	24	125	148	189	231	274	302	341	305	253	235	150	122	2675
IOWA BURLINGTON	19	148	165	217	241	284	315	353	327	270	243	175	147	2885
CHARLES CITY	22	137	157	190	226	258	285	336	290	241	207	130	115	2572
DES MOINES	30	155	170	203	236	276	303	346	299	263	227	156	136	2770
SIOUX CITY	30	164	177	216	254	300	320	363	320	270	236	160	146	2926
KAN. CONCORDIA	30	180	172	214	243	281	315	348	308	249	245	189	172	2916
DODGE CITY	30	205	191	249	268	305	335	359	335	290	266	218	198	3219
TOPEKA	18	159	160	193	215	260	287	310	304	263	229	173	149	2702
WICHITA	30	187	186	233	254	291	321	350	325	277	245	206	182	3057
KY. LOUISVILLE	30	115	135	188	221	283	303	324	295	256	219	148	114	2601
LA. NEW ORLEANS	30	160	158	213	247	292	287	260	269	241	260	200	157	2744
SHREVEPORT	19	151	172	214	240	298	332	339	322	289	273	208	177	3015
MAINE EASTPORT	22	133	151	196	201	245	248	275	260	205	175	105	115	2309
PORTLAND	30	155	174	213	226	268	286	312	294	229	202	146	148	2653
MD. BALTIMORE	30	148	170	211	229	270	295	299	272	238	212	164	145	2653
MASS. BLUE HILL OBS.	10	125	136	165	182	233	248	266	241	211	181	134	135	2257
BOSTON	30	148	168	212	222	263	283	300	280	232	207	152	148	2615
NANTUCKET	22	128	156	214	227	278	284	291	279	242	208	149	129	2585
MICH. ALPENA	24	86	124	198	228	261	303	339	285	204	159	70	67	2324
DETROIT	30	90	128	180	212	263	295	321	284	226	189	98	89	2375
LANSING	30	84	119	175	215	272	305	344	294	228	182	87	73	2378
ESCANABA	30	112	148	204	226	266	283	316	267	198	162	90	94	2366
GRAND RAPIDS	30	74	117	178	218	277	308	349	304	231	188	92	70	2406
MARQUETTE	30	78	113	172	207	248	268	305	251	186	142	68	66	2104
SAULT STE. MARIE	30	83	123	187	217	252	269	309	256	165	133	61	62	2117
MINN. DULUTH	30	125	163	221	235	268	282	328	277	203	166	100	107	2475
MINNEAPOLIS	30	140	166	200	231	272	302	343	296	237	193	115	112	2607
MISS. JACKSON	12	130	147	199	244	280	287	279	287	235	223	185	150	2646
VICKSBURG	30	136	141	199	232	284	304	291	297	254	244	183	140	2705
MO. COLUMBIA	30	147	164	207	232	281	296	341	298	262	225	166	138	2757
KANSAS CITY	30	154	170	211	235	278	313	347	308	266	235	178	151	2846
ST. JOSEPH	23	154	165	211	231	274	301	347	287	260	224	168	144	2766
ST. LOUIS	30	137	152	202	235	283	301	325	289	256	223	166	125	2694
SPRINGFIELD	30	145	164	213	238	278	305	342	310	269	233	183	140	2820
MONT. BILLINGS	21	140	154	208	236	283	301	372	332	258	213	136	129	2762
GREAT FALLS	19	154	176	245	261	299	299	381	342	256	206	132	133	2884
HAVRE	30	136	174	234	268	311	312	384	339	260	202	132	122	2874
HELENA	30	138	168	215	241	292	292	342	336	258	202	137	121	2742
MISSOULA	25	85	109	167	209	261	260	378	328	246	178	90	66	2377

STATE AND STATION	YEARS	JAN.	FEB.	MAR.	APR.	MAY	JUNE	JULY	AUG.	SEPT.	OCT.	NOV.	DEC.	ANNUAL
NEBR. LINCOLN	30	173	172	213	244	287	316	356	309	266	237	174	160	2907
NORTH PLATTE	30	181	179	221	246	282	310	343	304	264	242	184	169	2925
OMAHA	30	172	188	222	259	305	332	379	311	270	248	166	145	2997
VALENTINE	30	185	194	229	252	296	323	369	326	275	242	174	172	3037
NEV. ELY	22	186	197	262	260	300	354	359	344	303	255	204	187	3211
LAS VEGAS	8	239	251	314	336	386	411	383	364	345	301	258	250	3838
RENO	30	185	199	267	306	354	376	414	391	336	273	212	170	3483
WINNEMUCCA	30	142	155	207	255	312	346	395	375	316	242	177	139	3061
N. H. CONCORD	23	136	153	192	196	229	261	286	260	214	179	122	126	2354
MT. WASHINGTON OBS.	18	94	98	133	141	162	145	150	143	139	159	89	87	1540
N. J. ATLANTIC CITY	30	151	173	210	233	273	287	298	271	239	218	177	153	2683
TRENTON	30	145	168	203	235	277	294	309	273	239	208	160	142	2653
N. MEX. ALBUQUERQUE	30	221	218	273	299	343	365	340	317	299	279	245	219	3418
ROSWELL	21	218	223	286	306	330	333	341	313	266	266	242	216	3340
N. Y. ALBANY	30	125	151	194	213	266	301	317	286	224	192	115	112	2496
BINGHAMTON	30	94	119	151	170	226	256	266	230	184	158	92	79	2025
BUFFALO	30	110	125	180	212	274	319	338	297	239	183	97	84	2458
NEW YORK	30	154	171	213	237	268	289	302	271	235	213	169	155	2677
ROCHESTER	30	93	123	172	209	274	314	333	294	224	173	97	86	2392
SYRACUSE	30	87	115	165	197	261	295	316	276	211	163	81	74	2241
N. C. ASHEVILLE	30	146	161	211	247	289	292	268	250	235	222	179	146	2646
CAPE HATTERAS	9	152	168	206	259	293	301	286	265	214	202	169	154	2669
CHARLOTTE	30	165	177	230	267	313	316	291	277	247	243	198	167	2891
GREENSBORO	30	157	171	217	231	298	302	287	272	243	236	190	163	2767
RALEIGH	29	154	168	220	255	290	284	277	253	224	215	184	156	2680
WILMINGTON	30	179	180	237	279	314	312	286	273	237	238	206	178	2919
N. DAK. BISMARCK	30	141	170	205	236	279	294	358	307	243	198	130	125	2686
DEVILS LAKE	30	150	177	220	250	291	297	352	302	230	198	123	124	2714
FARGO	30	132	170	210	232	283	288	343	293	222	187	112	114	2586
WILLISTON	29	141	168	215	260	305	312	377	328	247	206	131	129	2819
OHIO CINCINNATI (ABBE)	30	115	137	186	222	273	309	323	295	253	205	138	118	2574
CLEVELAND	30	79	111	167	209	274	301	325	288	235	187	99	77	2352
COLUMBUS	30	112	132	177	215	270	296	323	291	250	210	131	101	2508
DAYTON	10	114	136	195	222	281	313	323	307	268	229	152	124	2664
SANDUSKY	30	100	128	183	229	285	312	343	302	248	201	111	91	2533
TOLEDO	30	93	120	170	203	263	296	331	298	241	196	106	92	2409
OKLA. OKLAHOMA CITY	29	175	182	235	253	290	329	352	331	282	243	201	175	3048
TULSA	18	152	164	200	213	244	287	314	308	281	241	207	172	2783
OREG. BAKER	22	118	143	198	251	302	313	406	368	289	215	132	100	2835
PORTLAND	30	77	97	142	203	246	249	329	275	218	134	87	65	2122
ROSEBURG	30	69	96	148	205	257	278	369	329	255	146	81	50	2283
PA. HARRISBURG	30	132	160	203	230	277	297	319	282	233	200	140	131	2604
PHILADELPHIA	30	142	166	203	231	270	281	288	253	225	205	158	142	2564
PITTSBURGH	25	89	114	163	200	239	260	283	250	234	180	114	76	2202
READING	30	133	151	195	220	259	275	293	259	219	198	144	127	2473
SCRANTON	30	108	138	178	199	251	269	290	249	213	183	120	105	2303
R. I. PROVIDENCE	30	145	168	211	221	271	285	292	267	226	207	153	143	2589
S. C. CHARLESTON	30	188	189	243	284	323	308	297	281	244	239	210	187	2993
COLUMBIA	30	173	183	233	274	312	312	291	283	243	242	202	166	2914
GREENVILLE	26	166	176	227	274	307	300	278	274	239	232	192	157	2822
S. DAK. HURON	30	153	177	213	250	295	321	367	320	260	212	142	134	2844
RAPID CITY	30	164	182	222	245	278	300	348	317	266	228	164	144	2858
TENN. CHATTANOOGA	30	126	146	187	239	290	295	278	266	247	220	169	128	2591
KNOXVILLE	30	124	144	189	237	281	288	277	248	237	213	157	120	2515
MEMPHIS	30	135	152	204	244	296	321	319	314	261	243	180	139	2808
NASHVILLE	30	123	142	196	241	285	308	292	279	250	224	168	126	2634
TEX. ABILENE	13	190	199	250	259	290	347	335	322	276	245	223	201	3137
AMARILLO	30	207	199	258	276	305	338	350	328	288	260	229	205	3243
AUSTIN	30	148	152	207	221	266	302	331	320	261	242	180	160	2790
BROWNSVILLE	30	147	152	187	210	272	297	326	311	246	252	165	151	2716
CORPUS CHRISTI	24	160	165	212	237	295	329	366	341	276	264	194	164	3003
DALLAS	30	155	159	220	238	279	326	341	325	274	240	191	163	2911
DEL RIO	27	173	173	230	237	259	279	331	319	252	240	195	178	2866
EL PASO	30	234	236	299	329	373	369	336	327	300	287	257	236	3583
GALVESTON	30	151	149	203	230	288	322	305	292	257	264	199	151	2811
HOUSTON	30	144	141	193	212	266	298	294	281	238	239	181	146	2633
PORT ARTHUR	30	153	149	209	235	292	317	285	281	252	256	191	148	2768
SAN ANTONIO	30	148	153	213	224	258	292	325	307	261	241	183	160	2765
UTAH SALT LAKE CITY	30	137	155	227	269	329	358	377	346	306	249	171	135	3059
VT. BURLINGTON	30	103	127	184	185	244	270	291	266	199	152	77	80	2178
VA. LYNCHBURG	26	153	169	216	243	288	297	288	264	235	217	177	158	2705
NORFOLK	30	156	174	223	257	304	311	296	282	237	220	182	161	2803
RICHMOND	30	144	166	211	248	280	296	286	263	230	211	176	152	2663
WASH. NORTH HEAD	22	76	97	135	182	221	214	226	186	170	123	87	66	1783
SEATTLE	30	74	99	154	201	247	234	304	248	197	122	77	62	2019
SPOKANE	30	78	120	197	262	308	309	397	350	264	177	86	57	2605
TATOOSH ISLAND	30	70	100	135	182	229	217	235	190	175	129	71	60	1793
WALLA WALLA	30	72	106	194	262	317	335	411	367	280	198	92	51	2685
W. VA. ELKINS	24	110	119	158	198	227	256	225	236	211	186	131	103	2160
PARKERSBURG	30	91	111	155	200	252	277	286	264	230	189	117	93	2265
WIS. GREEN BAY	30	121	148	194	210	251	279	314	266	213	176	110	106	2388
MADISON	30	126	147	196	214	258	285	336	288	230	198	116	108	2502
MILWAUKEE	30	116	134	191	218	267	293	340	292	235	193	125	106	2510
WYO. CHEYENNE	30	191	197	243	237	259	304	318	286	265	242	188	170	2900
LANDER	30	200	208	260	264	301	340	361	326	280	233	186	185	3144
SHERIDAN	30	160	179	226	245	286	303	367	333	266	221	153	145	2884
P. R. SAN JUAN	30	231	229	273	252	240	245	264	257	219	229	217	222	2878

4

Calculating Solar Radiation

The total solar radiation is the sum of direct, diffuse, and reflected radiation. At present, a statistical approach is the only reliable method of separating out the diffuse component of horizontal insolation. (The full detail of this method is contained in the article by Liu and Jordan cited below; their results are only summarized here.)

First ascertain the ratio of the daily insolation on a horizontal surface (measured at a particular weather station) to the extraterrestrial radiation on another horizontal surface (outside the atmosphere). This ratio (usually called the *percent of extraterrestrial radiation* or percent ETR) can be determined from the National Weather Records Center; it is also given in the article by Liu and Jordan. With a knowledge of the percent ETR, you can use the accompanying graph to determine the percentage of diffuse radiation on a horizontal surface. For example, 50 percent ETR corresponds to 38 percent diffuse radiation and 62 percent direct radiation.

You are now prepared to convert the direct and diffuse components of the horizontal insolation into the daily total insolation on south-facing tilted or vertical surfaces. The conversion factor for the direct component (F_D), depends on the latitude (L), the tilt angle of the surface (represented by the Greek letter Beta, or β),

and the sunset hour angles (represented by the Greek letter Omega, or ω), of the horizontal and tiled surfaces:

horizontal surface: $\cos \omega$
$$= -\tan L \tan \delta$$
tilted surface: $\cos \omega'$
$$= -\tan (L - \beta)\tan \delta$$

where the declination δ is found from the graph in Appendix 1 and $\beta = 90°$ applies to vertical surfaces. Depending on the value of these two angles ω and ω', the calculation of F_D is slightly different. If ω is less the ω', then:

$$F_D = \frac{\cos(L - \beta)}{\cos L} \times \frac{\sin \omega - \omega \cos \omega'}{\sin \omega - \omega \cos \omega}$$

If ω' is smaller than ω, then:

$$F_D = \frac{\cos (L - \beta)}{\cos L} \times \frac{\sin \omega' - \omega' \cos \omega'}{\sin \omega - \omega \cos \omega}$$

The direct component of the radiation on a tilted or vertical surface is $I'_D = F_D (I_D)$, where I_D is the direct horizontal insolation

The treatment of diffuse and reflected radiation is a bit different. The diffuse radiation is assumed to come uniformly from all corners of the sky, so one need only determine the fraction of the sky exposed to a tilted surface and reduce the horizontal diffuse radiation accordingly. The diffuse radiation on a surface tilted at an angle β is:

$$I'_d = 1 + \cos \beta/2(I_d)$$

where I_d is the daily horizontal diffuse radiation.

The reflected radiation on a tilted surface is:

$$I'_d = \rho(1 - \cos \beta/2)(I_D + I_d)$$

where ρ (the Greek letter Rho) is the reflectance of the horizontal surface. (Source: Liu, B.Y.H. and R.C. Jordan "Availability of Solar Energy for Flat-Plate Solar Heat Collectors." in *Low Temperature Engineering Applications of Solar Energy*, edited by Richard C. Jordan, New York: ASHRAE, 1967.)

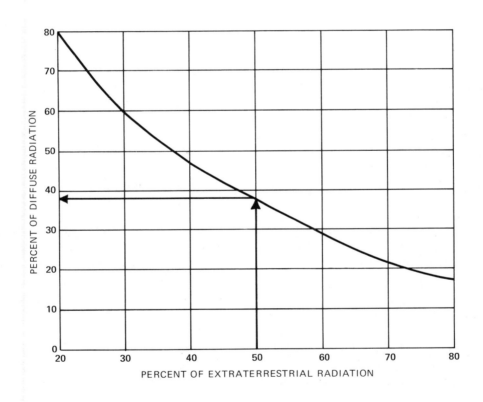

5
Degree Days and Design Temperatures

The hourly, monthly, and yearly heat losses from a house depend on the temperature difference between the indoor and outdoor air, as explained earlier. To aid in the calculation of these heat losses, ASHRAE publishes the expected winter design temperatures and the monthly and yearly total degree days for many cities and towns in the United States.

The maximum heat loss rate occurs when the temperature is lowest, and you need some idea of the lowest likely temperature in your locale in order to size a conventional heating unit. The ASHRAE *Handbook of Fundamentals* provides three choices—the "median of annual extremes" and the "99%" and "97 1/2%" design temperatures. The median of annual extremes is the average of the lowest winter temperatures recorded in each of the past 30 to 40 years. The "99%" and "97 1/2%" design temperatures are the temperatures which are normally exceeded during those percentages of the time in December, January and February. We list the "97 1/2%" temperatures here together with the average winter temperatures. For example, the temperature will fall below 19°F for 2 1/2 per-

cent of the time (about 2 days) during a typical Birmingham winter. Consult the ASHRAE *Handbook of Fundamentals* for more detailed listings.

Degree days gauge heating requirements over the long run. One degree day occurs for every day the average outdoor temperature is 1°F below 65°F, which is the base for degree day calculations because most houses don't require any heating until outdoor temperatures fall below this level. For example, if the outdoor air temperature remained constant at 30°F for the entire month of January, then $31(65 - 30) = 1085$ degree days would result. Both monthly and yearly total degree days are listed in these tables, but only the months from September to May are included here because very little heating is needed in summer. The yearly total degree days are the sum over all 12 months. More complete listings of monthly and yearly degree days can be found in the ASHRAE *Guide and Data Book* or *Handbook of Fundamentals*. (Sources: ASHRAE, *Guide and Data Book*, 1970; ASHRAE, *Handbook of Fundamentals, 1981*, reprinted by permission.)

Top table

State	City	Avg. Winter Temp	Design Temp	Sep	Oct	Nov	Dec	Jan	Feb	Mar	Apr	May	Yearly Total
	Tampa	66.4	36	0	0	60	171	202	148	102	0	0	683
	West Palm Beach	68.4	40	0	0	6	65	87	64	31	0	0	253
Ga.	Athens	51.8	17	12	115	405	632	642	529	431	141	22	2929
	Atlanta	51.7	18	18	124	417	648	636	518	428	147	25	2961
	Augusta	54.5	20	0	78	333	552	549	445	350	90	0	2397
	Columbus	54.8	23	0	87	333	543	557	434	338	96	0	2383
	Macon	56.2	23	0	71	297	502	557	403	295	63	0	2136
	Rome	49.9	16	24	161	474	701	710	577	468	177	34	3326
	Savannah	57.8	24	0	47	246	437	437	353	254	45	0	1819
Hawaii	Hilo	71.9	59	0	0	0	0	0	0	0	0	0	0
	Honolulu	74.2	60	0	0	0	0	0	0	0	0	0	0
Idaho	Boise	39.7	4	132	415	792	1017	1113	854	722	438	245	5809
	Lewiston	41.0	6	123	403	756	1063	1063	815	694	426	239	5542
	Pocatello	34.8	-8	172	493	900	1166	1324	1058	905	555	319	7033
Ill.	Chicago	37.5	-4	81	326	753	1113	1209	1044	890	480	211	6155
	Moline	36.4	-7	99	335	774	1181	1314	1100	918	450	189	6408
	Peoria	38.1	-2	87	326	759	1113	1218	1025	849	426	183	6025
	Rockford	34.8	-7	114	400	837	1221	1333	1137	961	516	236	6830
	Springfield	40.6	-1	72	291	696	1023	1135	935	769	354	136	5429
Ind.	Evansville	45.0	6	66	220	606	896	955	767	620	237	68	4435
	Fort Wayne	37.3	0	105	378	783	1135	1178	1028	890	471	189	6205
	Indianapolis	39.6	0	90	316	723	1051	1113	949	809	432	177	5699
	South Bend	36.6	-2	111	372	777	1125	1221	1070	933	525	239	6439
Iowa	Burlington	37.6	-4	93	322	768	1135	1259	1042	859	426	177	6114
	Des Moines	35.5	-7	96	363	828	1225	1370	1137	915	438	180	6588
	Dubuque	32.7	-11	156	450	906	1287	1420	1204	1026	546	260	7376
	Sioux City	34.0	-10	108	369	867	1240	1435	1198	989	483	214	6951
	Waterloo	32.6	-12	138	428	909	1296	1460	1221	1023	531	229	7320
Kans.	Dodge City	42.5	3	33	251	666	939	1051	840	719	354	124	4986
	Goodland	37.8	-2	81	381	810	1073	1166	955	884	507	236	6141
	Topeka	41.7	3	57	270	672	980	1122	893	722	330	124	5182
	Wichita	44.2	5	33	229	618	905	1023	804	645	270	87	4620
Ky.	Covington	41.4	3	75	291	669	983	1035	893	756	390	149	5265
	Lexington	43.8	6	54	239	609	902	946	818	685	325	105	4683
	Louisville	44.0	8	54	248	609	890	930	818	682	315	105	4660
La.	Alexandria	57.5	25	0	56	273	431	471	361	260	69	0	1921
	Baton Rouge	59.8	25	0	31	216	369	409	294	208	33	0	1560
	Lake Charles	60.5	29	0	19	210	341	381	274	195	39	0	1459
	New Orleans	61.0	32	0	19	192	322	363	258	192	39	0	1385
	Shreveport	56.2	22	0	47	297	477	552	426	304	81	0	2184
Me.	Caribou	24.4	-18	336	682	1044	1535	1690	1470	1308	858	468	9767
	Portland	33.0	-5	195	508	807	1215	1339	1182	1042	675	372	7511
Md.	Baltimore	43.7	12	48	264	585	905	936	820	679	327	90	4654
	Frederick	42.0	7	66	307	624	955	995	876	741	384	127	5087
Mass.	Boston	40.0	6	60	316	603	983	1088	972	846	513	208	5634
	Pittsfield	32.6	-5	219	524	831	1231	1196	1063	1063	660	326	7578
	Worcester	34.7	-3	147	450	774	1172	1271	1123	998	612	304	6969

Bottom table

State	City	Avg. Winter Temp	Design Temp	Sep	Oct	Nov	Dec	Jan	Feb	Mar	Apr	May	Yearly Total
Ala.	Birmingham	54.2	19	6	93	363	555	592	462	363	108	9	2551
	Huntsville	51.3	13	12	127	426	663	694	557	434	138	19	3070
	Mobile	59.9	26	0	22	213	357	415	300	211	42	0	1560
	Montgomery	55.4	22	0	68	330	527	543	417	316	90	0	2291
Alaska	Anchorage	23.0	-25	516	930	1284	1572	1631	1316	1293	879	592	10864
	Fairbanks	6.7	-51	642	1203	1833	2254	2359	1901	1739	1068	555	14279
	Juneau	32.1	-7	483	725	921	1135	1237	1070	1073	810	601	9075
	Nome	13.1	-32	693	1094	1455	1820	1879	1666	1770	1314	930	14171
Ariz.	Flagstaff	35.6	0	201	558	867	1073	1169	991	911	651	437	7152
	Phoenix	58.5	31	0	22	234	415	474	328	217	75	0	1765
	Tucson	58.1	29	0	25	231	406	471	344	242	75	6	1800
	Winslow	43.0	9	6	245	711	1008	1054	770	601	291	96	4782
	Yuma	64.2	37	0	0	108	264	307	190	90	15	0	974
Ark.	Fort Smith	50.3	9	12	127	450	704	781	596	456	144	22	3292
	Little Rock	50.5	19	9	127	465	716	756	577	434	126	9	3219
	Texarkana	54.2	22	0	78	345	561	626	468	350	105	0	2533
Calif.	Bakersfield	55.4	31	0	37	282	502	546	364	267	105	19	2122
	Burbank	58.6	36	6	43	177	301	366	277	239	138	81	1646
	Eureka	49.9	32	258	329	414	499	546	470	505	438	372	4643
	Fresno	53.3	28	0	84	354	577	605	426	335	162	62	2611
	Long Beach	57.8	36	9	47	171	316	397	311	264	171	93	1803
	Los Angeles	57.4	41	42	78	180	291	372	302	288	219	158	2061
	Oakland	53.5	35	45	127	309	481	527	400	353	255	180	2870
	Sacramento	53.9	30	21	56	321	546	583	414	332	178	72	2502
	San Diego	59.5	42	21	43	135	236	298	235	214	135	90	1458
	San Francisco	55.1	42	102	118	231	388	443	336	319	279	239	3001
	Santa Maria	54.3	32	96	146	270	391	459	370	363	282	233	2967
Colo.	Alamosa	29.7	-17	279	639	1065	1420	1476	1162	1020	696	440	8529
	Colorado Springs	37.3	-1	132	456	825	1032	1128	938	893	582	319	6423
	Denver	37.6	-2	117	428	819	1035	1132	938	887	558	288	6283
	Grand Junction	39.3	-8	30	313	786	1113	1209	907	729	387	146	5641
	Pueblo	40.4	-5	54	326	750	986	1085	871	772	429	174	5462
Conn.	Bridgeport	39.9	4	66	307	615	986	1079	966	853	510	208	5617
	Hartford	37.3	1	117	394	714	1101	1190	1042	908	519	205	6235
	New Haven	39.0	5	87	347	648	1011	1097	991	871	543	245	5897
Del.	Wilmington	42.5	12	51	270	588	927	980	874	735	387	112	4930
D. C.	Washington	45.7	16	33	217	519	834	871	762	626	288	74	4224
Fla.	Daytona Beach	64.5	32	0	0	75	211	248	190	140	15	0	879
	Fort Myers	68.6	38	0	0	24	109	146	101	62	0	0	442
	Jacksonville	61.9	29	0	12	144	310	332	246	174	21	0	1239
	Key West	73.1	55	0	0	0	28	40	31	9	0	0	108
	Lakeland	66.7	35	0	0	57	164	195	146	99	0	0	661
	Miami	71.1	44	0	0	0	65	74	56	19	0	0	214
	Miami Beach	72.5	45	0	0	0	40	56	36	9	0	0	141
	Orlando	65.7	33	0	0	72	198	220	165	105	6	0	766
	Pensacola	60.4	29	0	19	195	353	400	277	183	36	0	1463
	Tallahassee	60.1	25	0	28	198	360	375	286	202	36	0	1485

Left table (Mich. – N.M.)

State	City	Avg. Winter Temp	Design Temp	Sep	Oct	Nov	Dec	Jan	Feb	Mar	Apr	May	Yearly Total
Mich.	Alpena	29.7	-5	273	580	912	1268	1404	1299	1218	777	446	8506
	Detroit	37.2	4	87	360	738	1088	1181	1058	936	522	220	6232
	Escanaba	29.6	-7	243	539	924	1293	1445	1296	1203	777	456	8481
	Flint	33.1	-1	159	465	843	1212	1330	1198	1066	639	319	7377
	Grand Rapids	34.9	2	135	434	804	1147	1259	1134	1011	579	279	6894
	Lansing	34.8	2	138	431	813	1163	1262	1142	1011	579	273	6909
	Marquette	30.2	-8	240	527	936	1268	1411	1268	1187	771	468	8393
	Muskegon	36.0	4	120	400	762	1088	1209	1100	995	594	310	6696
	Sault Ste. Marie	27.7	-12	279	580	951	1367	1525	1380	1277	810	477	9048
Minn.	Duluth	23.4	-19	330	632	1131	1581	1745	1518	1355	840	490	10000
	Minneapolis	28.3	-14	189	505	1014	1454	1631	1380	1166	621	288	8382
	Rochester	28.8	-17	186	474	1005	1438	1593	1366	1150	630	301	8295
Miss.	Jackson	55.7	21	0	65	315	502	546	414	310	87	0	2239
	Meridian	55.4	20	0	81	339	518	543	417	310	81	0	2289
	Vicksburg	56.9	23	0	53	279	462	512	384	282	69	0	2041
Mo.	Columbia	42.3	2	54	251	651	967	1076	874	716	324	121	5046
	Kansas City	43.9	4	39	220	612	905	1032	818	682	294	109	4711
	St. Joseph	40.3	-1	60	285	708	1039	1172	949	769	348	133	5484
	St. Louis	43.1	4	60	251	627	936	1026	848	704	312	121	4900
	Springfield	44.5	5	45	223	600	877	973	781	660	291	105	4900
Mont.	Billings	34.5	-10	186	487	897	1135	1296	1100	970	570	285	7049
	Glasgow	26.4	-25	270	608	1104	1466	1711	1439	1187	648	335	8996
	Great Falls	32.8	-20	258	543	921	1169	1349	1154	1063	642	384	7750
	Havre	28.1	-22	306	595	1065	1367	1584	1364	1181	657	338	8700
	Helena	31.1	-17	294	601	1002	1265	1438	1170	1042	651	381	8129
	Kalispell	31.4	-7	321	654	1020	1240	1401	1134	1029	639	397	8191
	Miles City	31.2	-19	174	502	972	1296	1504	1252	1057	579	276	7723
	Missoula	31.5	-7	303	651	1035	1287	1420	1120	970	621	391	8125
Neb.	Grand Island	36.0	-6	108	381	834	1172	1314	1089	908	462	211	6530
	Lincoln	38.8	-4	75	301	726	1066	1237	1016	834	402	171	5864
	Norfolk	34.0	-11	111	397	873	1234	1414	1179	983	498	233	6979
	North Platte	35.5	-6	123	440	885	1166	1271	1039	930	519	248	6684
	Omaha	35.6	-5	105	357	828	1175	1355	1126	939	465	208	6612
	Scottsbluff	35.9	-8	138	459	876	1128	1231	1008	921	552	285	6673
Nev.	Elko	34.0	-13	225	561	924	1197	1314	1036	911	621	409	7433
	Ely	33.1	-6	234	592	939	1184	1308	1075	977	672	456	7733
	Las Vegas	53.5	23	0	78	387	617	688	487	335	111	6	2709
	Reno	39.3	2	204	490	801	1026	1073	823	729	510	357	6332
	Winnemucca	36.7	1	210	536	876	1091	1172	916	837	573	363	6761
N. H.	Concord	33.0	-11	177	505	822	1240	1358	1184	1032	636	298	7383
N. J.	Atlantic City	43.2	14	39	251	549	880	936	848	741	420	133	4812
	Newark	42.8	11	30	248	573	921	983	876	729	381	118	4589
	Trenton	42.4	12	57	264	576	924	989	885	753	399	121	4980
N. M.	Albuquerque	45.0	14	12	229	642	868	930	703	595	288	81	4348
	Raton	38.1	-2	126	431	825	1048	1116	904	834	543	301	6228
	Roswell	47.5	16	18	202	573	840	930	641	481	201	31	3793
	Silver City	48.0	14	6	183	525	729	791	605	518	261	87	3705

Right table (N.Y. – S.D.)

State	City	Avg. Winter Temp	Design Temp	Sep	Oct	Nov	Dec	Jan	Feb	Mar	Apr	May	Yearly Total
N. Y.	Albany	34.6	-5	138	440	777	1194	1311	1156	992	564	239	6875
	Binghamton	36.6	-2	141	406	732	1107	1190	1081	949	543	229	6451
	Buffalo	34.5	3	141	440	777	1156	1256	1145	1039	645	329	7062
	New York	42.8	11	30	233	540	902	986	885	760	408	118	4871
	Rochester	35.4	2	126	415	747	1125	1234	1123	1014	597	279	6748
	Schenectady	35.4	-5	123	422	756	1159	1283	1159	970	543	211	6650
	Syracuse	35.2	-2	132	415	744	1153	1271	1140	1004	570	248	6756
N. C.	Asheville	46.7	13	48	245	555	775	784	683	592	273	87	4042
	Charlotte	50.4	18	6	124	438	691	691	582	481	156	22	3191
	Greensboro	47.5	14	33	192	513	778	784	672	552	234	47	3805
	Raleigh	49.4	16	21	164	450	716	725	616	487	180	34	3393
	Wilmington	54.6	23	0	74	291	521	546	462	357	96	0	2347
	Winston-Salem	48.4	14	21	171	483	747	753	652	524	207	37	3595
N. D.	Bismarck	26.6	-24	222	577	1083	1463	1708	1442	1203	645	329	8851
	Devils Lake	22.4	-23	273	642	1191	1634	1872	1579	1345	690	381	9901
	Fargo	24.8	-22	219	574	1107	1569	1789	1520	1262	732	332	9226
	Williston	25.2	-21	261	601	1122	1513	1758	1473	1262	681	357	9243
Ohio	Akron-Canton	38.1	1	96	381	726	1070	1138	1016	871	489	202	6037
	Cincinnati	45.1	8	39	208	558	862	915	790	642	294	96	4410
	Cleveland	37.2	2	105	384	738	1088	1159	1047	918	552	260	6351
	Columbus	39.7	2	84	347	714	1039	1088	949	809	426	171	5660
	Dayton	39.8	0	78	310	696	1045	1097	955	809	429	167	5622
	Mansfield	36.9	1	114	397	768	1110	1169	1042	924	543	245	6403
	Toledo	36.4	1	117	406	792	1138	1200	1056	924	543	242	6494
	Youngstown	36.8	1	120	412	771	1104	1169	1047	921	540	248	6417
Okla.	Oklahoma City	48.3	11	15	164	498	766	868	664	527	189	34	3725
	Tulsa	47.7	12	18	158	522	787	893	683	539	213	47	3860
Ore.	Astoria	45.6	27	210	375	561	679	753	622	636	480	363	5186
	Eugene	45.6	22	129	366	585	719	803	627	589	426	279	4726
	Medford	43.2	21	78	372	678	871	918	697	642	432	242	5008
	Pendleton	42.6	3	111	350	711	884	1017	773	617	396	205	5127
	Portland	45.6	21	114	335	597	735	825	644	586	396	245	4635
	Roseburg	46.3	25	105	329	567	713	766	608	570	405	267	4491
	Salem	45.4	21	111	338	594	729	822	647	611	417	273	4754
Pa.	Allentown	38.9	3	90	353	693	1045	1116	1002	849	471	167	5810
	Erie	36.8	7	102	391	714	1063	1169	1081	973	585	288	6451
	Harrisburg	41.2	9	63	298	648	992	1045	907	766	396	124	5251
	Philadelphia	41.8	11	60	297	620	965	1016	889	747	392	118	5144
	Pittsburgh	38.4	5	105	375	726	1063	1119	1002	874	480	195	5987
	Reading	42.4	6	54	257	597	939	1001	885	735	372	105	4945
	Scranton	37.2	2	132	434	762	1104	1156	1028	893	498	195	6254
	Williamsport	38.5	1	111	375	717	1073	1122	1002	856	468	177	5934
R. I.	Providence	38.8	6	96	372	660	1023	1110	988	868	534	236	5954
S. C.	Charleston	57.9	26	0	34	210	425	443	367	273	42	0	1794
	Columbia	54.0	20	0	84	345	577	570	470	357	81	0	2484
	Florence	54.5	21	0	78	315	552	552	459	347	84	0	2387
	Greenville-Spartanburg	51.6	18	6	121	399	651	660	546	446	132	19	2980
S. D.	Huron	28.8	-16	165	508	1014	1432	1628	1355	1125	600	288	8223

State	City	Avg. Winter Temp	Design Temp	Sep	Oct	Nov	Dec	Jan	Feb	Mar	Apr	May	Yearly Total
	Rapid City	33.4	− 9	165	481	897	1172	1333	1145	1051	615	326	7345
	Sioux Falls	30.6	−14	168	462	972	1361	1544	1285	1082	573	270	7839
Tenn.	Bristol	46.2	11	51	236	573	828	828	700	598	261	68	4143
	Chattanooga	50.3	15	18	143	468	698	722	577	453	150	25	3254
	Knoxville	49.2	13	30	171	489	725	732	613	493	198	43	3494
	Memphis	50.5	17	18	130	447	698	729	585	456	147	22	3232
	Nashville	48.9	12	30	158	495	732	778	644	512	189	40	3578
Tex.	Abilene	53.9	17	0	99	366	586	642	470	347	114	0	2624
	Amarillo	47.0	8	18	205	570	797	877	664	546	252	56	3985
	Austin	59.1	25	0	31	225	388	468	325	223	51	0	1711
	Corpus Christi	64.6	32	0	0	120	220	291	174	109	0	0	914
	Dallas	55.3	19	0	62	321	524	601	440	319	90	6	2363
	El Paso	52.9	21	0	84	414	648	685	445	319	105	0	2700
	Galveston	62.2	32	0	6	147	276	360	263	189	33	0	1274
	Houston	61.0	28	0	6	183	307	384	288	192	36	0	1396
	Laredo	66.0	32	0	0	105	217	267	134	74	0	0	797
	Lubbock	48.8	11	18	174	513	744	800	613	484	201	31	3578
	Port Arthur	60.5	29	0	22	207	329	384	274	192	39	0	1447
	San Antonio	60.1	25	0	31	204	363	428	286	195	39	0	1546
	Waco	57.2	21	0	43	270	456	536	389	270	66	0	2030
	Wichita Falls	53.0	15	0	99	381	632	698	518	378	120	6	2832
Utah	Milford	36.5	− 1	99	443	867	1141	1252	988	822	519	279	6497
	Salt Lake City	38.4	5	81	419	849	1082	1172	910	763	459	233	6052
Vt.	Burlington	29.4	−12	207	539	891	1349	1513	1333	1187	714	353	8269
Va.	Lynchburg	46.0	15	51	223	540	822	849	731	605	267	78	4166
	Norfolk	49.2	20	0	136	408	698	738	655	533	216	37	3421
	Richmond	47.3	14	36	214	495	784	815	703	546	219	53	3865
	Roanoke	46.1	15	51	229	549	825	834	722	614	261	65	4150
Wash.	Olympia	44.2	21	198	422	636	753	834	675	645	450	307	5236
	Seattle	46.9	28	129	329	543	657	738	599	577	396	242	4424
	Spokane	36.5	− 2	168	493	879	1082	1231	980	834	531	288	6655
	Walla Walla	43.8	12	87	310	681	843	986	745	589	342	177	4805
	Yakima	39.1	6	144	450	828	1039	1163	868	713	435	220	5941
W. Va.	Charleston	44.8	9	63	254	591	865	880	770	648	300	96	4476
	Elkins	40.1	1	135	400	729	992	1008	896	791	444	198	5675
	Huntington	45.0	10	63	257	585	856	880	764	636	294	99	4446
	Parkersburg	43.5	8	60	264	606	905	942	826	691	339	115	4754
Wisc.	Green Bay	30.3	−12	174	484	924	1333	1494	1313	1141	654	335	8029
	La Crosse	31.5	−12	153	437	924	1339	1504	1277	1070	540	245	7589
	Madison	30.9	− 9	174	474	930	1330	1473	1274	1113	618	310	7863
	Milwaukee	32.6	− 6	174	471	876	1252	1376	1193	1054	642	372	7635
Wyo.	Casper	33.4	−11	192	524	942	1169	1290	1084	1020	657	381	7410
	Cheyenne	34.2	− 6	219	543	909	1085	1212	1042	1026	702	428	7381
	Lander	31.4	−16	204	555	1020	1299	1417	1145	1017	654	381	7870
	Sheridan	32.5	−12	219	539	948	1200	1355	1154	1051	642	366	7680

6

Insulating Values of Materials

The conduction heat flow through a wall, window, door, roof, ceiling, or floor decreases as more *resistance* is placed in the path of the flow. All materials have some resistance to conduction heat flow. Those that have high resistance are called *insulators*; those with low resistance are called *conductors*.

Insulators are compared to one another according to their R-values, which are a measure of their resistance. The R-value of a material increases with its thickness—a 2-inch thick sheet of polystyrene has twice the resistance of a 1-inch sheet. And two similar building materials that differ in density will also differ in R-value. Generally, though not always, the lighter material will have a higher R-value because it has more pockets of air trapped in it. Finally, the average temperature of a material also affects its R-value. The colder it gets, the better most materials retard the flow of heat.

Knowledge of the R-values of insulators and other components permits us to calculate the heat transmission through a wall or other building surface. Toward this end, we list the R-values of many common buliding materials in the first table. R-values are given per inch of thickness and for standard thicknesses. If you have some odd size not listed in the table, use the R-value per inch thickness and multiply by its thickness. Unless otherwise noted, the R-values are quoted for a temperature of 75°F.

Further tables list R-values for surface air films and for air spaces, both of which have insulating values. These R-values vary markedly with the reflectance of the surfaces facing the air film or space. Radiation heat flow is very slow across an air space with aluminum foil on one side, for example, and the R-value of such an air space is correspondingly high. This is why fiberglass batt insulation is often coated with an aluminized surface. In the tables we have used three categories of surface: non-reflective (such as painted wood or metal), fairly-reflective (such as aluminum-coated paper), and highly-reflective (such as metallic foil). The R-value of an air film or air space also depends on the orientation of the surface and the direction of heat flow that we are trying to retard. These differences are reflected in the tables.

The total resistance R_t of a wall or other building surface is just the sum of the R-values of all its components—including air films and spaces. The coefficient of heat transmission, or U-value, is the inverse of the total resistance ($U = 1/R_t$). To get the rate of heat loss through a wall, for example, you multiply its U-value by the total surface area of the wall and by the temperature difference between the indoor and outdoor air. The next table in this section lists U values for windows and skylights. Here, again,

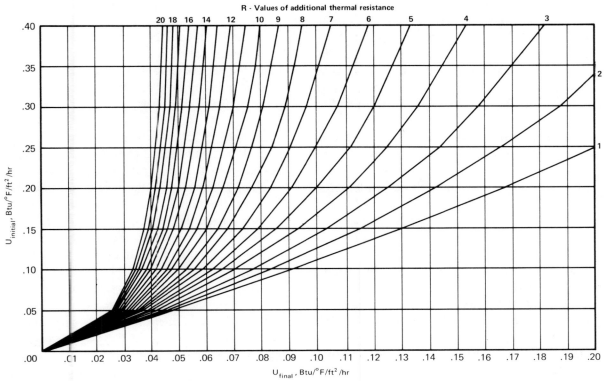

R - Values of additional thermal resistance

NOTES: Final U value obtained by adding an amount of thermal resistance, R, to an exterior skin having a coefficient of heat transmission, $U_{initial}$.

the U-value depends upon the surface orientation, the direction of heat flow, and the season of the year. The U-values in these tables apply only to the glazing surfaces; to include the effects of a wood sash, multiply these U-values by about 80 to 90 percent, depending upon the area of the wood.

For the avid reader seeking more detailed information about the insulating values of building materials, we recommend the ASHRAE *Handbook of Fundamentals*, from which most of the present data were taken. Tables there list resistances and conductances of many more materials than are given here. Sample calculations of the U-values of typical frame and masonry walls, roofs, and floors are also provided there.

Recently hard coat, low-e glass, a new kind of low-emissivity (low-e) coated glazing, became available. It stands up better than soft coat low-e glass and shares many of soft coat's energy conserving properties. U-values are shown

below for hard coat, soft coat and suspended film, low-e glass

Glazing	Air Space	U-Values Winter	U-Values Summer
Double			
soft coat	1/4"	0.44	0.48
	1/2"	0.32	0.32
hard coat	1/4"	0.52	0.54
	1/2"	0.40	0.44
Triple			
soft coat	1/4"	0.32	0.37
	1/2"	0.23	0.37
suspended	3/8"	0.25	0.28
film	1/4"	0.31	0.35
	1/2"	0.23	0.25

(Data from the Sealed Insulated Glass Manufacturers Association.)

In all of this discussion, no mention has been made of the relative costs of all the various building alternatives. To a large extent, these depend upon the local building materials suppliers. But charts in the next two sections of this appendix will help you to assess the savings in fuel costs that can be expected from adding insulation.

By adding insulating materials to a wall or other building surface, you can lower its U-value, or heat transmission coefficient. But it takes a much greater amount of insulation to lower a small U-value than it does to lower a large U-value. For example, adding 2 inches of polyurethane insulation ($R_t = 12$) to a solid 8-inch concrete wall reduces the U-value from 0.66 to 0.07, or almost a factor of 10. Adding the same insulation to a good exterior stud wall reduces the U-value from 0.069 to 0.038, or less than a factor of 2. Mathematically, if U_i is the inital U-value of a building surface, and R is the resistance of the added insulation, the final U-value (U_f) is:

$$U_f = U_i/1 + RU_i$$

If you don't have a pocket calculator handy, the following chart will help you to tell at a glance the effects of adding insulation to a wall or other building surface. The example shows you how to use this chart.

Example If 3 1/2 inches of fiberglass insulation (R = 11) is added to an uninsulated stud wall having a U-value of 0.23, what is the final U-value? When adding the insulation, you remove the insulating value of the air space (R = 1.01) inside the wall, so the net increase in resistance is R = 10. To use the chart, begin at $U_i = 0.23$ on the left-hand scale. Move horizontally to intersect the curve numbered R = 10. Drop down from this point to the bottom scale to find the final U-value, $U_f = 0.069$. With this information, you can now use the Heat Conduction Cost Chart in the next section to find the fuel savings resulting from the added insulation.

Thermal Properties of Typical Building and Insulating Materials—Design Values[a]

Description	Density lb/ft^3	Resistance [c] (R) Per inch thickness $(1/\lambda)$ h•ft^2•F/Btu	Resistance [c] (R) For thickness listed $(1/C)$ h•ft^2•F/Btu
BUILDING BOARD			
Boards, Panels, Subflooring, Sheathing			
Woodboard Panel Products			
Asbestos-cement board......................	120	0.25	—
Asbestos-cement board...................0.125 in.	120	—	0.03
Asbestos-cement board...................0.25 in.	120	—	0.06
Gypsum or plaster board..............0.375 in.	50	—	0.32
Gypsum or plaster board................0.5 in.	50	—	0.45
Gypsum or plaster board.............0.625 in.	50	—	0.56
Plywood (Douglas Fir)[o]................	34	1.25	—
Plywood (Douglas Fir) ....0.25 in.	34	—	0.31
Plywood (Douglas Fir) ....0.375 in.	34	—	0.47
Plywood (Douglas Fir) ....0.5 in.	34	—	0.62
Plywood (Douglas Fir) ....0.625 in.	34	—	0.77
Plywood or wood panels ....0.75 in.	34	—	0.93
Vegetable Fiber Board			
Sheathing, regular density ....0.5 in.	18	—	1.32
............0.78125 in.	18	—	2.06
Sheathing intermediate density ....0.5 in.	22	—	1.22
Nail-base sheathing......................0.5 in.	25	—	1.14
Shingle backer ....0.375 in.	18	—	0.94
Shingle backer ....0.3125 in.	18	—	0.78
Sound deadening board...................0.5 in.	15	—	1.35
Tile and lay-in panels, plain or			
acoustic................................	18	2.50	—
..................................0.5 in.	18	—	1.25
..................................0.75 in.	18	—	1.89
Laminated paperboard	30	2.00	—
Homogeneous board from			
repulped paper...........................	30	2.00	—
Hardboard			
Medium density...........................	50	1.37	—
High density, service temp. service			
underlay	55	1.22	—
High density, std. tempered...................	63	1.00	—
Particleboard			
Low density.............................	37	1.85	—
Medium density...........................	50	1.06	—
High density	62.5	0.85	—
Underlayment ....0.625 in.	40	—	0.82
Wood subfloor ....0.75 in.		—	0.94
BUILDING MEMBRANE			
Vapor—permeable felt	—	—	0.06
Vapor—seal, 2 layers of mopped			
15-lb felt	—	—	0.12
Vapor—seal, plastic film......................	—		Negl.
FINISH FLOORING MATERIALS			
Carpet and fibrous pad........................	—	—	2.08
Carpet and rubber pad	—	—	1.23
Cork tile ....0.125 in.	—	—	0.28
Terrazzo ....1 in.	—	—	0.08
Tile—asphalt, linoleum, vinyl, rubber.............	—	—	0.05
vinyl asbestos			
ceramic..................................			
Wood, hardwood finish ....0.75 in.	—	—	0.68
INSULATING MATERIALS			
Blanket and Batt[d]			
Mineral Fiber, fibrous form processed			
from rock, slag, or glass			
approx.[e] 3–4 in.........................	0.3–2.0	—	11[d]
approx.[e] 3.5 in.........................	0.3–2.0	—	13[d]
approx.[e] 5.5–6.5 in......................	0.3–2.0	—	19[d]
approx.[e] 6–7.5 in.	0.3–2.0	—	22[d]
approx.[e] 9–10 in.........................	0.3–2.0	—	30[d]
approx.[e] 12–13 in........................	0.3–2.0	—	38[d]

Thermal Properties of Typical Building and Insulating Materials—Design Values[a]

Description	Density lb/ft³	Resistance [c] (R)	
		Per inch thickness $(1/\lambda)$ h•ft²• F/Btu	For thickness listed $(1/C)$ h•ft²• F/Btu
Board and Slabs			
Cellular glass	8.5	2.86	—
Glass fiber, organic bonded	4–9	4.00	—
Expanded perlite, organic bonded................	1.0	2.78	—
Expanded rubber (rigid)	4.5	4.55	—
Expanded polystyrene extruded			
Cut cell surface	1.8	4.00	—
Smooth skin surface	1.8–3.5	5.00	—
Expanded polystyrene, molded beads	1.0	3.85	—
	1.25	4.00	—
	1.5	4.17	—
	1.75	4.17	—
	2.0	4.35	—
Cellular polyurethane[f] (R-11 exp.)(unfaced)........	1.5	6.25	—
Cellular polyisocyanurate[n] (R-11 exp.) (foil faced, glass fiber-reinforced core)	2.0	7.20	—
Nominal 0.5 in................................		—	3.6
Nominal 1.0 in...............................		—	7.2
Nominal 2.0 in...............................		—	14.4
Mineral fiber with resin binder	15.0	3.45	
Mineral fiberboard, wet felted			
Core or roof insulation	16–17	2.94	—
Acoustical tile.............................	18.0	2.86	—
Acoustical tile.............................	21.0	2.70	—
Mineral fiberboard, wet molded			
Acoustical tile[h]	23.0	2.38	—
Wood or cane fiberboard			
Acoustical tile[g] ....0.5 in.	—	—	1.25
Acoustical tile[g] ....0.75 in.	—	—	1.89
Interior finish (plank, tile)......................	15.0	2.86	—
Cement fiber slabs (shredded wood with Portland cement binder	25–27.0	2.0–1.89	—
Cement fiber slabs (shredded wood with magnesia oxysulfide binder).................	22.0	1.75	—
LOOSE FILL			
Cellulosic insulation (milled paper or wood pulp)	2.3–3.2	3.70–3.13	—
Sawdust or shavings	8.0–15.0	2.22	—
Wood fiber, softwoods.........................	2.0–3.5	3.33	—
Perlite, expanded	2.0–4.1	3.7–3.3	—
	4.1–7.4	3.3–2.8	—
	7.4–11.0	2.8–2.4	—
Mineral fiber (rock, slag or glass)			
approx.[e] 3.75–5 in........................	0.6–2.0		11.0
approx.[e] 6.5–8.75 in.	0.6–2.0		19.0
approx.[e] 7.5–10 in.	0.6–2.0		22.0
approx.[e] 10.25–13.75 in.	0.6–2.0		30.0
Mineral fiber (rock, slag or glass)			
approx.[e] 3.5 in. (closed sidewall application)	2.0–3.5	—	12.0-14.0
Vermiculite, exfoliated	7.0–8.2	2.13	—
	4.0–6.0	2.27	—
FIELD APPLIED			
Polyurethane foam	1.5-2.5	6.25-5.26	—
Ureaformaldehyde foam.......................	0.7-1.6	3.57-4.55	—
Spray cellulosic fiber base	2.0-6.0	3.33-4.17	—
PLASTERING MATERIALS			
Cement plaster, sand aggregate..................	116	0.20	—
Sand aggregate.........................0.375 in.	—	—	0.08
Sand aggregate ....0.75 in.	—	—	0.15
Gypsum plaster:			
Lightweight aggregate ....0.5 in.	45	—	0.32
Lightweight aggregate ....0.625 in.	45	—	0.39
Lightweight agg. on metal lath............0.75 in.	—	—	0.47
Perlite aggregate	45	0.67	—

Thermal Properties of Typical Building and Insulating Materials—Design Values[a]

Description	Density lb/ft³	Resistance [c] (R)	
		Per inch thickness $(1/\lambda)$ h·ft²·F/Btu	For thickness listed $(1/C)$ h·ft²·F/Btu
PLASTERING MATERIALS			
Sand aggregate	105	0.18	—
Sand aggregate .... 0.5 in.	105	—	0.09
Sand aggregate .... 0.625 in.	105	—	0.11
Sand aggregate on metal lath .... 0.75 in.	—	—	0.13
Vermiculite aggregate	45	0.59	—
MASONRY MATERIALS			
Concretes			
Cement mortar	116	0.20	—
Gypsum-fiber concrete 87.5% gypsum, 12.5% wood chips	51	0.60	—
Lightweight aggregates including ex-	120	0.19	—
panded shale, clay or slate; expanded	100	0.28	—
slags; cinders; pumice; vermiculite;	80	0.40	—
also cellular concretes	60	0.59	—
	40	0.86	—
	30	1.11	—
	20	1.43	—
Perlite, expanded	40	1.08	—
	30	1.41	—
	20	2.00	—
Sand and gravel or stone aggregate (oven dried)	140	0.11	—
Sand and gravel or stone aggregate (not dried)	140	0.08	—
Stucco	116	0.20	—
MASONRY UNITS			
Brick, common[i]	120	0.20	—
Brick, face[i]	130	0.11	—
Clay tile, hollow:			
1 cell deep .... 3 in.	—	—	0.80
1 cell deep .... 4 in.	—	—	1.11
2 cells deep .... 6 in.	—	—	1.52
2 cells deep .... 8 in.	—	—	1.85
2 cells deep .... 10 in.	—	—	2.22
3 cells deep .... 12 in.	—	—	2.50
Concrete blocks, three oval core:			
Sand and gravel aggregate .... 4 in.	—	—	0.71
.... 8 in.	—	—	1.11
.... 12 in.	—	—	1.28
Cinder aggregate .... 3 in.	—	—	0.86
.... 4 in.	—	—	1.11
.... 8 in.	—	—	1.72
.... 12 in.	—	—	1.89
Lightweight aggregate .... 3 in.	—	—	1.27
(expanded shale, clay, slate .... 4 in.	—	—	1.50
or slag; pumice): .... 8 in.	—	—	2.00
.... 12 in.	—	—	2.27
Concrete blocks, rectangular core.[j,k]			
Sand and gravel aggregate			
2 core, 8 in. 36 lb.	—	—	1.04
Same with filled cores[l]	—	—	1.93
Lightweight aggregate (expanded shale, clay, slate or slag, pumice):			
3 core, 6 in. 19 lb.	—	—	1.65
Same with filled cores[l]	—	—	2.99
2 core, 8 in. 24 lb.	—	—	2.18
Same with filled cores[l]	—	—	5.03
3 core, 12 in. 38 lb.	—	—	2.48
Same with filled cores[l]	—	—	5.82
Stone, lime or sand	—	0.08	—
Gypsum partition tile:			
3 • 12 • 30 in. solid	—	—	1.26
3 • 12 • 30 in. 4-cell	—	—	1.35
4 • 12 • 30 in. 3-cell	—	—	1.67

Thermal Properties of Typical Building and Insulating Materials—Design Values[a]

Description	Density lb/ft^3	Resistance[c] (R) Per inch thickness ($1/\lambda$) h·ft^2·F/Btu	Resistance[c] (R) For thickness listed ($1/C$) h·ft^2·F/Btu
ROOFING[h]			
Asbestos-cement shingles	120	—	0.21
Asphalt roll roofing	70	—	0.15
Asphalt shingles	70	—	0.44
Built-up roofing .... 0.375 in.	70	—	0.33
Slate .... 0.5 in.	—	—	0.05
Wood shingles, plain and plastic film faced	—	—	0.94
SIDING MATERIALS (on flat surface)			
Shingles			
Asbestos-cement	120	—	0.21
Wood, 16 in., 7.5 exposure	—	—	0.87
Wood, double, 16-in., 12-in. exposure	—	—	1.19
Wood, plus insul. backer board, 0.3125 in.	—	—	1.40
Siding			
Asbestos-cement, 0.25 in., lapped	—	—	0.21
Asphalt roll siding	—	—	0.15
Asphalt insulating siding (0.5 in. bed.)	—	—	1.46
Hardboard siding, 0.4375 in.	40	0.67	
Wood, drop, 1 · 8 in.	—	—	0.79
Wood, bevel, 0.5 · 8 in., lapped	—	—	0.81
Wood, bevel, 0.75 · 10 in., lapped	—	—	1.05
Wood, plywood, 0.375 in., lapped	—	—	0.59
Aluminum or Steel[m], over sheathing			
Hollow-backed	—	—	0.61
Insulating-board backed nominal 0.375 in.	—	—	1.82
Insulating-board backed nominal 0.375 in., foil backed			2.96
Architectural glass	—	—	0.10
WOODS (12% Moisture Content)[o,p]			
Hardwoods			
Oak	41.2-46.8	0.89-0.80	—
Birch	42.6-45.4	0.87-0.82	—
Maple	39.8-44.0	0.94-0.88	—
Ash	38.4-41.9	0.94-0.88	—
Softwoods			
Southern Pine	35.6-41.2	1.00-0.89	—
Douglas Fir-Larch	33.5-36.3	1.06-0.99	—
Southern Cypress	31.4-32.1	1.11-1.09	—
Hem-Fir, Spruce-Pine-Fir	24.5-31.4	1.35-1.11	—
West Coast Woods, Cedars	21.7-31.4	1.48-1.11	—
California Redwood	24.5-28.0	1.35-1.22	—

[a] Except where otherwise noted, all values are for a mean temperature of 75 F. Representative values for dry materials, selected by ASHRAE TC 4.4, are intended as design (not specification) values for materials in normal use. Insulation materials in actual service may have thermal values that vary from design values depending on their in-situ properties (e.g., density and moisture content). For properties of a particular product, use the value supplied by the manufacturer or by unbiased tests.

[b] To obtain thermal conductivities in But/h·ft^2·F, divide the λ value by 12 in./ft.

[c] Resistance values are the reciprocals of C before rounding off C to two decimal places.

[d] Does not include paper backing and facing, if any. Where insulation forms a boundary (reflective or otherwise) of an air space, see Tables 2A and 2B for the insulating value of an air space with the appropriate effective emittance and temperature conditions of the space.

[e] Conductivity varies with fiber diameter. (See Chapter 20, Thermal Conductivity section.) Insulation is produced in different densities, therefore, there is a wide variation in thickness for the same R-value among manufacturers. No effort should be made to relate any specific R-value to any specific density or thickness.

[f] Values are for aged, unfaced, board stock. For change in conductivity with age of expanded urethane, see Chapter 20, Factors Affecting Thermal Conductivity.

[g] Insulating values of acoustical tile vary, depending on density of the board and on type, size and depth of perforations.

[h] ASTM C 855-77 recognizes the specification of roof insulation on the basis of the C-values shown. Roof insulation is made in thickness to meet these values.

[i] Face brick and common brick do not always have these specific densities. When density differs from that shown, there will be a change in thermal conductivity.

[j] At 45 F mean temperature. Data on rectangular core concrete blocks differ from the above data on oval core blocks, due to core configuration, different mean temperatures, and possibly differences in unit weights. Weight data on the oval core blocks tested are not available.

[k] Weights of units approximately 7.625 in. high and 15.75 in. long. These weights are given as a means of describing the blocks tested, but conductance values are all for 1 ft^2 of area.

[l] Vermiculite, perlite, or mineral wool insulation. Where insulation is used, vapor barriers or other precautions must be considered to keep insulation dry.

[m] Values for metal siding applied over flat surfaces vary widely, depending on amount of ventilation of air space beneath the siding; whether air space is reflective or nonreflective; and on thickness, type, and application of insulating backing-board used. Values given are averages for use as design guides, and were obtained from several guarded hotbox tests (ASTM C236) or calibrated hotbox (ASTM C 976) on hollow-backed types and types made using backing-boards of wood fiber, foamed plastic, and glass fiber. Departures of ±50% or more from the values given may occur.

[n] Time-aged values for board stock with gas-barrier quality (0.001 in. thickness or greater) aluminum foil facers on tow major surfaces.

[o] See Ref. 5.

[p] See Ref. 6, 7, 8 and 9. The conductivity values listed are for heat transfer across the grain. The thermal conductivity of wood varies linearly with the density and the density ranges listed are those normally found for the wood species given. If the density of the wood species is not known, use the mean conductivity value.

R-VALUES OF AIR SPACES				
Orientation & Thickness of Air Space	Direction of Heat Flow	R-value for Air Space Facing:[‡]		
		Non-reflective surface	Fairly reflective surface	Highly reflective surface
Horizontal ¾"	up*	0.87	1.71	2.23
4"		0.94	1.99	2.73
¾"	up[†]	0.76	1.63	2.26
4"		0.80	1.87	2.75
¾"	down*	1.02	2.39	3.55
1½"		1.14	3.21	5.74
4"		1.23	4.02	8.94
¾	down[†]	0.84	2.08	3.25
1½"		0.93	2.76	5.24
4"		0.99	3.38	8.03
45° slope ¾"	up*	0.94	2.02	2.78
4"		0.96	2.13	3.00
¾"	up[†]	0.81	1.90	2.81
4"		0.82	1.98	3.00
¾"	down*	1.02	2.40	3.57
4"		1.08	2.75	4.41
¾"	down[†]	0.84	2.09	3.34
4"		0.90	2.50	4.36
Vertical ¾"	across*	1.01	2.36	3.48
4"		1.01	2.34	3.45
¾"	across[†]	0.84	2.10	3.28
4"		0.91	2.16	3.44

[‡]One side of the air space is a non-reflective surface.
*Winter conditions.
[†]Summer conditions.
SOURCE: ASHRAE, *Handbook of Fundamentals*, 1972. Reprinted by permission.

R-VALUES OF AIR FILMS				
Type and Orientation of Air Film	Direction of Heat Flow	R-value for Air Film On:		
		Non-reflective surface	Fairly reflective surface	Highly reflective surface
Still air:				
Horizontal	up	0.61	1.10	1.32
Horizontal	down	0.92	2.70	4.55
45° slope	up	0.62	1.14	1.37
45° slope	down	0.76	1.67	2.22
Vertical	across	0.68	1.35	1.70
Moving air:				
15 mph wind	any*	0.17	—	—
7½ mph wind	any[†]	0.25	—	—

*Winter conditions.
[†]Summer conditions.
SOURCE: ASHRAE, *Handbook of Fundamentals*, 1972. Reprinted by permission.

U-VALUES OF WINDOWS AND SKYLIGHTS		
Description	U-values[1]	
	Winter	Summer
Vertical panels:		
Single pane flat glass	1.13	1.06
Insulating glass—double[2]		
3/16" air space	0.69	0.64
1/4" air space	0.65	0.61
1/2" air space	0.58	0.56
Insulating glass—triple[2]		
1/4" air spaces	0.47	0.45
1/2" air spaces	0.36	0.35
Storm windows		
1-4" air space	0.56	0.54
Glass blocks[3]		
6 × 6 × 4" thick	0.60	0.57
8 × 8 × 4" thick	0.56	0.54
' same, with cavity divider	0.48	0.46
Single plastic sheet	1.09	1.00
Horizontal panels:[4]		
Single pane flat glass	1.22	0.83
Insulating glass—double[2]		
3/16" air space	0.75	0.49
1/4" air space	0.70	0.46
1/2" air space	0.66	0.44
Glass blocks[3]		
11 × 11 × 3" thick, with cavity divider	0.53	0.35
12 × 12 × 4" thick, with cavity divider	0.51	0.34
Plastic bubbles[5]		
single-walled	1.15	0.80
double-walled	0.70	0.46

[1] in units of $Btu/hr/ft^2/°F$

[2] double and triple refer to the number of lights of glass.

[3] nominal dimensions.

[4] U-values for horizontal panels are for heat flow *up* in winter and *down* in summer.

[5] based on area of opening, not surface.

SOURCE: ASHRAE, *Handbook of Fundamentals,* 1972. Reprinted by permission.

7
Heat Conduction Cost Chart

The Heat Conduction Cost Chart provided here simplifies the calculation of total seasonal heat loss through wall, roof, or floor. It also facilitates the study of alternative constructions and possible savings due to added insulation. An example of the use of the chart is included:

- For a building surface with a U-value of 0.58, start at point (1).
- Follow up the oblique line to the horizontal line representing the total heating degree days for the location, in this case 7,000 degree days (2),
- Move vertically from this point to find a heat loss of 95,000 Btu/ft^2 per season for the surface (3).
- Continue vertically to the oblique line representing the total area of the surface, 100 square feet (4).
- Moving horizontally from this point, the total heat loss through the entire surface for the season is 9,500,000 Btu (5).
- Continue horizontally to the oblique line representing the cost per million Btu of heat energy, in this case $9 per million Btu (6).
- Moving vertically down from this point, the total cost for the season of the heat through that surface is $86 (7).

The lower right graph converts the apparent cost to a "real cost of energy" through the use of a multiplication factor. This factor might reflect:

(1) Estimated future cost of energy—design decisions based on present energy costs make little sense as costs soar.

(2) Real environmental cost of using fossil fuels—this particularly includes pollution and the depletion of natural resources, both directly as fuels burn and indirectly as they are brought to the consumer from the source.

(3) Initial investment cost—use of the proper multiplication factor would give the quantity of increased investment made possible by resultant yearly fuel savings.

For example, heat costs may increase by a factor of 10. Continue down from the last point until you intersect the oblique line representing the multiplication factor 10 (8). Then move horizontally left to arrive at the adjusted seasonal heating cost through the building surface of $862 (9).

The numerical values of the chart can be changed by a factor of ten. For example, to determine the heat transfer through a really good exterior wall, U = 0.05, use U = 0.5 on the chart and divide the final answer by ten. Each of the graphs can be used independently of one another. For example, knowing a quantity of energy and its price, the upper right graph gives the total cost of that energy.

The Heat Conduction Cost Chart can help you compare the energy costs of two different methods of insulation. For example, an insulated stud wall has a U-value of 0.07 and an uninsulated wall has a U-value of 0.23. The *difference* 0.23 − 0.07 = 0.16 can be run through the chart in the same way as done for a single U-value. Assuming 5000 degree days, 100 square feet of wall, and $9 per million Btu, the savings in heating costs for one year is about $21, or more than the cost of insulation.

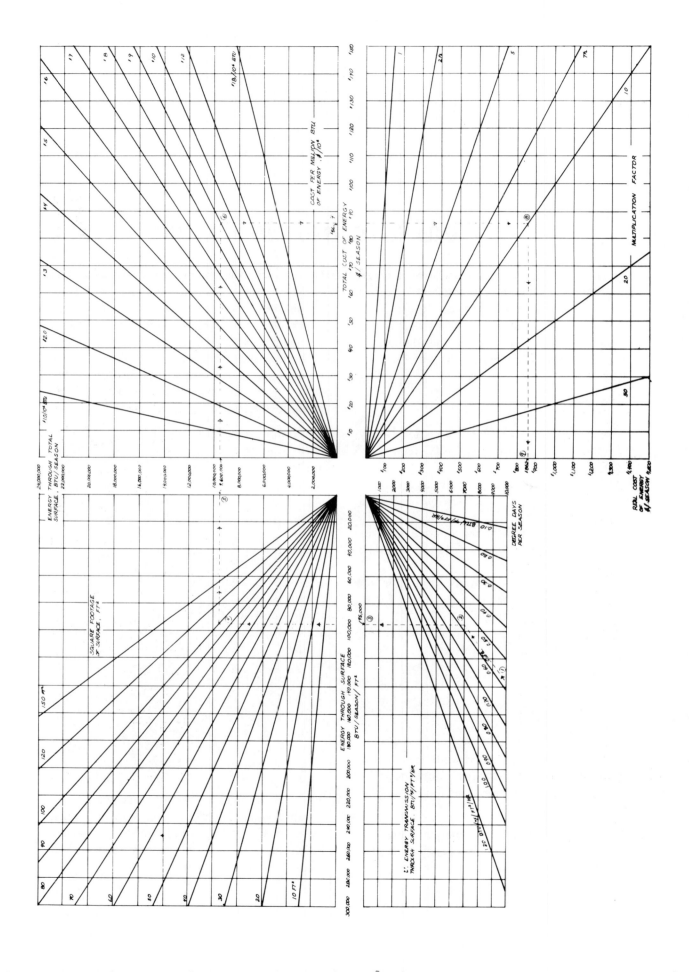

8

Air Infiltration Cost Chart

The use of this chart to calculate the costs of heat loss through air infiltration is similar to the use of the Heat Conduction Cost Chart. An example of its use is included:

- Start at point (1) for an air infiltration rate of 45 ft^3/(hr ft).
- Follow up the oblique line to the horizontal line representing the total heating degree days for the location, in this case 7000 degree days (2).
- Move vertically from this point to find that 146,000 Btu are consumed each heating season per crack foot (3).
- Continue vertically to the oblique line corresponding to the total crack length, in this case 30 feet (4).
- Move horizontally from this point to the total seasonal heat loss through the window crack— 4,400,000 Btu (5).
- Continue horizontally to the oblique line representing the cost per million Btu of heat energy, in this case $6 per million Btu (6).
- Move vertically down from this point to the total cost for heat lost through the crack during an entire heating season, or $26.75.

As with the Heat Conduction Cost Chart, the bottom right graph permits a conversion of this apparent cost to a "real cost of energy" through the use of a multiplication factor. In this example, a factor of 10 is used. Continue down from the last point until you intersect the oblique line representing a multiplication factor 10 (8). Then move horizontally left to arrive at an adjusted heating cost of about $270 per heating season (9).

As with the previous chart, you can use this Air Infiltration Cost Chart to make quick evaluations of the savings resulting from changes in the rate of air infiltratrion. For example, if a wood-sash, double-hung window is weatherstripped, the air infiltration rate will drop from 39 to 24 cubic feet per hour per crack foot. By moving through the chart from a starting point of 15 ft^3/(hr ft), you arrive ultimately at the savings resulting from weatherstripping. Assuming 5000 degree days, 15 feet of crack, and $9 per million Btu, we get an immediate savings of about $6 in the first heating season. Since weatherstripping costs a few cents per foot, it can pay for itself in fuel savings within a few weeks.

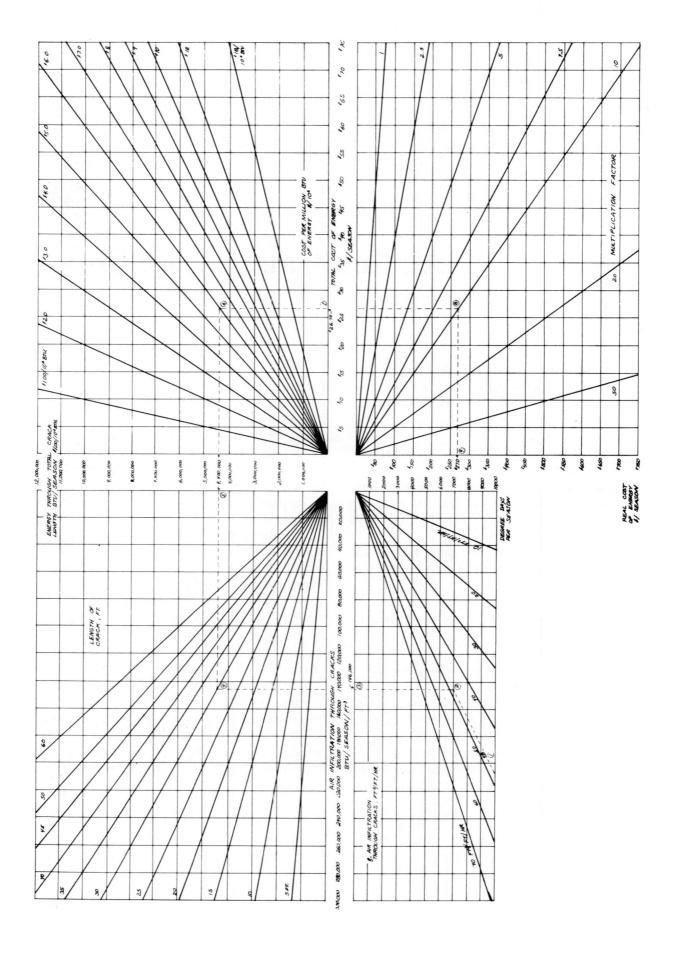

9

Emittances and Absorptions of Materials

Radiation is an important method of heat transfer between two surfaces. As explained earlier, any warm body emits energy in the form of electromagnetic radiation. We might say every warm object has an "aura" that is invisible to our eyes.

Sunlight is one form of electromagnetic radiation and thermal radiation is another. They differ only in wavelength. Sunlight comes in wavelengths ranging from 0.3 to 3.0 microns. The wavelengths of thermal radiation from warm bodies (say 100°F) range from 3 to 50 microns

When radiation strikes a surface of any material, it is either absorbed, reflected, or transmitted. Each material absorbs, reflects, and transmits radiation differently—according to its physical and chemical characteristics and the wavelength of the incoming radiation. For example, glass transmits most of the sunlight hitting it but absorbs almost all thermal radiation.

We can assign numerical ratings that gauge the percentage of radiation absorbed, reflected, or transmitted by a material. These numbers depend upon the temperature of the material and the wavelength of the radiation. We usually define the absorptance (represented by the Greek letter Alpha, or α) of a material as the ratio of solar energy (in the wavelength range 0.3 to 3.0 microns) absorbed to the total solar energy incident:

$$\alpha = I_a/I = \text{absorbed solar energy/incident solar energy}$$

The *reflectance* (represented by the Greek letter Rho, or ρ) and *transmittance* (represented by the Greek letter Tau, or τ) are similarly defined ratios:

$$\rho = I_r/I = \text{reflected solar energy/incident solar energy}$$

$$\tau = I_t/I = \text{transmitted solar energy/incident solar energy}$$

Because all the sunlight is either absorbed, reflected, or transmitted, $\alpha + \rho + \tau = 1$. For opaque solids, no energy is transmitted, so that $\alpha + \rho = 1$ or $\rho = 1 - \alpha$. If we know the absorptance of an opaque material, we also know its reflectance.

Once absorbed, this radiant energy is transformed into heat energy—the motion of molecules. The body becomes warmer and emits more radiation of its own. The emittance (represented by the Greek letter Epsilon, or ϵ of a material is a numerical indicator of that material's propensity to radiate away its energy. The emittance is defined as the ratio of the thermal radiation emitted from a material to the thermal radiation emitted by a hypothetical "black-body" with the same shape and temperature:

$$\epsilon = R_m/R_b$$
$$= \text{radiation from material/}$$
$$\text{radiation from blackbody}$$

With $\epsilon = 1$, the blackbody is a theoretically "perfect" emitter of thermal radiation.

A knowledge of the absorptances and emittances of materials helps us to evaluate their relative thermal performance. For example, brick, masonry and concrete have emittances around 0.9—so they are better heat radiators than galvanized iron, which has an emittance between 0.13 and 0.28. With an absorptance greater than 0.9, asphalt paving absorbs much more of the sunlight than sand ($\alpha = 0.60$ to 0.75), as any-one who has walked barefoot from parking lot to beach can testify

The ratio α/ϵ of the absorptance (of short-wave solar radiation) to the emittance (of long-wave thermal radiation) has special importance in the design of solar collectors. In general, you want materials with high values of α/ϵ for the absorber coating. Then a large percentage of solar radiation is absorbed, but only a small amount lost by re-radiation. Materials with high values of both α and α/ϵ are called "selective surfaces."

The ensuing tables list absorptances and emittances of many common and some uncommon materials. They are grouped into two categories according to whether α/ϵ is less than or greater than 1.0.

CLASS I SUBSTANCES: Absorptance to Emittance Ratios (α/ε) (Continued) Less than 1.0

Substance	α	ε	α/ε
Small hole in large box, furnace or enclosure	0.99	0.99	1.0
"Hohlraum," theoretically perfect black body	1.00	1.0	1.0

CLASS II SUBSTANCES: Absorptance to Emittance Ratios (α/ε) Greater than 1.0

Substance	α	ε	α/ε
Black silk velvet	0.99	0.97	1.02
Alfalfa, dark green	0.97	0.95	1.02
Lamp black	0.98	0.95	1.03
Black paint on aluminum	0.94-0.98	0.88	1.07-1.11
Granite	0.55	0.44	1.25
Dull brass, copper, lead	0.2-0.4	0.4-0.65	1.63-2.0
Graphite	0.78	0.41	1.90
Stainless steel wire mesh	0.63-0.86	0.23-0.28	2.70-3.0
Galvanized sheet iron, oxidized	0.80	0.28	2.86
Galvanized iron, clean, new	0.65	0.13	5.00
Aluminum foil	0.15	0.05	3.00
Cobalt oxide on polished nickel*	0.93-0.94	0.24-0.40	3.9
Magnesium	0.30	0.07	4.3
Chromium	0.49	0.08	6.13
Nickel black on galvanized iron*	0.89	0.12	7.42
Cupric oxide on sheet aluminum*	0.85	0.11	7.73
Nickel black on polished nickel*	0.91-0.94	0.11	8.27-8.55
Polished zinc	0.46	0.02	23.0

*Selective surfaces
SOURCES: ASHRAE, *Handbook of Fundamentals,* 1972.
Bowden, *Alternative Sources of Energy,* July 1973.
Duffie and Beckman, *Solar Energy Thermal Processes,* 1974.
McAdams, *Heat Transmission,* 1954.
Severns and Fellows, *Air Conditioning and Refrigeration,* 1966.
Sounders, *The Engineer's Companion,* 1966.

CLASS I SUBSTANCES: Absorptance to Emittance Ratios (α/ε) Less than 1.0

Substance	α	ε	α/ε
White plaster	0.07	0.91	0.08
Snow, fine particles, fresh	0.13	0.82	0.16
White paint on aluminum	0.20	0.91	0.22
Whitewash on galvanized iron	0.22	0.90	0.24
White paper	0.25-0.28	0.95	0.26-0.29
White enamel on iron	0.25-0.45	0.90	0.28-0.50
Ice, with sparse snow cover	0.31	0.96-0.97	0.32
Snow, ice granules	0.33	0.89	0.37
Aluminum oil base paint	0.45	0.90	0.50
Asbestos felt	0.25	0.50	0.50
White powdered sand	0.45	0.84	0.54
Green oil base paint	0.50	0.90	0.56
Bricks, red	0.55	0.92	0.60
Asbestos cement board, white	0.59	0.96	0.61
Marble, polished	0.5-0.6	0.90	0.61
Rough concrete	0.60	0.97	0.62
Concrete	0.60	0.88	0.68
Grass, wet	0.67	0.98	0.68
Grass, dry	0.67-0.69	0.90	0.76
Vegetable fields and shrubs, wilted	0.70	0.90	0.78
Oak leaves	0.71-0.78	0.91-0.95	0.78-0.82
Grey paint	0.75	0.95	0.79
Desert surface	0.75	0.90	0.83
Common vegetable fields and shrubs	0.72-0.76	0.90	0.82
Red oil base paint	0.74	0.90	0.82
Asbestos, slate	0.81	0.96	0.84
Ground, dry plowed	0.75-0.80	0.70-0.96	0.83-0.89
Linoleum, red-brown	0.84	0.92	0.91
Dry sand	0.82	0.90	0.91
Green roll roofing	0.88	0.91-0.97	0.93
Slate, dark grey	0.89	—	0.95
Bare moist ground	0.90	0.95	0.95
Wet sand	0.91	0.95	0.96
Water	0.94	0.95-0.96	0.98
Black tar paper	0.93	0.93	1.0
Black gloss paint	0.90	0.90	1.0

10
Specific Heats and
Heat Capacities of Materials

Different materials abosorb different amounts of heat while undergoing the same temperature rise. For example, ten pounds of water will absorb 100 Btu during a 10°F temperature rise, but 10 pounds of cast iron will absorb only 12 Btu over the same range. There are two common measures of the ability of material to absorb and store heat—its specific heat and its heat capacity.

The specific heat of a material is the number of Btu absorbed by a pound of that material as its temperture rises 1°F. All specific heats vary with temperature and a distinction must be made between the true and the mean specific heat. The true specific heat is the number of Btu ab-

sorbed per pound per °F temperature rise at a fixed temperature. Over a wider temperature range, the mean specific heat is the average number of Btu's absorbed per pound per °F temperature rise. In the following table only true specific heats are given—for room temperature unless otherwise noted.

The heat capacity of a material is the amount of heat absorbed by one cubic foot of that material during a 1°F temperature rise. The heat capacity is just the product of the density of the material (in lb/ft^3) times its specific heat (Btu/(lb°F)). Specific heats, heat capacities, and densities of common building materials and other substances are given in the following table.

Material	Specific Heat (Btu/lb/°F)	Density (lb/ft^3)	Heat Capacity (Btu/ft^3/°F)
Air (at 1 atmosphere)	0.24 [75]	0.075	0.018
Aluminum (alloy 1100)	0.214	171	36.6
Asbestos fiber	0.25	150	37.5
Asbestos insulation	0.20	36	7.2
Ashes, wood	0.20	40	8.0
Asphalt	0.22	132	29.0
Bakelite	0.35	81	28.4
Brick, building	0.2	123	24.6
Brass, red (85% Cu, 15% Zn)	0.09	548	49.3
Brass, yellow (65% Cu, 35% Zn)	0.09	519	46.7
Bronze	0.104	530	55.1
Cellulose	0.32	3.4	1.1
Cement (Portland clinker)	0.16	120	19.2
Chalk	0.215	143	30.8
Charcoal (wood)	0.20	15	3.0

Specific Heats and Heat Capacities of Materials

Material	Specific Heat (Btu/lb/°F)	Density (lb/ft³)	Heat Capacity (Btu/ft³/°F)
Clay	0.22	63	13.9
Coal	0.3	90	27.0
Concrete (stone)	0.22	144	31.7
Copper (electrolytic)	0.092	556	51.2
Cork (granulated)	0.485	5.4	2.6
Cotton (fiber)	0.319	95	30.3
Ethyl alcohol	0.68	49.3	33.5
Fireclay brick	0.198 [212]	112	22.2
Glass, crown (soda-lime)	0.18	154	27.7
Glass, flint (lead)	0.117	267	31.2
Glass, pyrex	0.20	139	27.8
Glass, "wool"	0.157	3.25	0.5
Gypsum	0.259	78	20.2
Hemp (fiber)	0.323	93	30.0
Ice	0.487 [32]	57.5	28.0
Iron, cast	0.12 [212]	450	54.0
Lead	0.031	707	21.8
Limestone	0.217	103	22.4
Magnesium	0.241	108	26.0
Marble	0.21	162	34.0
Nickel	0.105	555	58.3
Octane	0.51	43.9	22.4
Paper	0.32	58	18.6
Paraffin	0.69	56	38.6
Porcelain	0.18	162	29.2
Rock salt	0.219	136	29.8
Salt water	0.75	72	54.0
Sand	0.191	94.6	18.1
Silica	0.316	140	44.2
Silver	0.056	654	36.6
Steel (mild)	0.12	489	58.7
Stone (quarried)	0.2	95	19.0
Tin	0.056	455	25.5
Tungsten	0.032	1210	38.7
Water	1.0 [39]	62.4	62.4
Wood, white oak	0.570	47	26.8
Wood, white fir	0.65	27	17.6
Wood, white pine	0.67	27	18.1
Zinc	0.092	445	40.9

*Values are for room temperature unless otherwise noted in brackets.

11
Metric/English Equivalents and Conversion Factors

METRIC / ENGLISH EQUIVALENTS

English Measure	Metric Equivalent	Metric Measure	English Equivalent
inch	2.54 centimeters	millimeter	0.04 inch
foot	30.50 centimeters	centimeter	0.39 inch
yard	0.91 meter	meter	3.28 feet
mile (statute)	1.60 kilometers	meter	1.09 yards
		kilometer	0.62 miles
square inch	6.45 square centimeters		
square foot	929.00 square centimeters	square centimeter	0.16 square inch
square yard	0.84 square meter	square meter	1.19 square yards
square mile	2.60 square kilometers	square kilometer	0.38 square mile
ounce	28.30 grams	gram	.035 ounces
pound (mass)	0.45 kilogram	kilogram	2.20 pounds
short ton	907.00 kilograms	ton (1,000 kg)	1.10 short tons
fluid ounce	29.60 milliliters	milliliters	0.03 fluid ounce
pint	0.47 liter	liter	1.06 quarts
quart	0.95 liter	liter	0.26 gallon
gallon	3.78 liters	cubic meter	35.3 cubic feet
cubic foot	0.03 cubic meter	cubic meter	1.3 cubic yards
cubic yard	0.76 cubic meter		
Btu	251.98 calories	calorie	0.004 Btu
pound (force)	4.45 newtons	newton	0.225 pound (force)

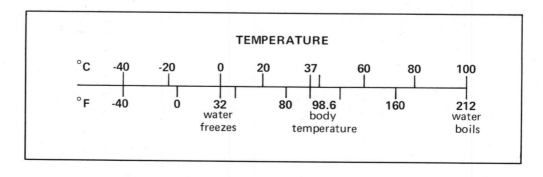

Multiply:	By:	To obtain:
Acres	43,560	Square feet
Acre-feet	1,233.5	Cubic meters
Barrels, oil (crude)	5.8×10^6	Btu
Barrels, oil	5.615	Cubic feet
Btu	777.48	Foot-pounds
Btu	1,055	Joules
Btu	0.29305	Watt-hours
Btu/hr/ft²/°F	5.682×10^4	Watts/cm²/°C
Btu per square foot	0.271	Langleys (cal/cm²)
Calories	3.9685×10^{-3}	Btu
Calories	4.184	Joules
Cords	128	Cubic feet
Cubic feet	0.037037	Cubic yards
Cubic feet	7.48	Gallons
Cubic feet per second	448.83	Gallons per minute
Feet of water (39.2°F)	0.4335	Pounds per square inch
Feet of water	0.88265	Inches of mercury at 32°F
Gallons	0.1337	Cubic feet
Gallons of water at 60°F	8.3453	Pounds
Horsepower	33,000	Foot-pounds per minute
Horsepower	42.42	Btu per minute
Horsepower	2,546	Btu per hour
Horsepower	1.014	Metric horsepower

Multiply:	By:	To obtain:
Horsepower	0.7457	Kilowatts
Inches of mercury at 32°F	0.4912	Pounds per square inch
Kilowatts	56.90	Btu per minute
Kilowatts	1.341	Horsepower
Kilowatt-hours	3,413	Btu
Kilowatt-hours	2.66×10^6	Foot-pounds
Langleys (cal/cm²)	3.69	Btu per square foot
Langleys per minute	0.0698	Watts per square centimeter
Microns	1×10^{-4}	Centimeters
Months (mean calendar)	730.1	Hours
Newtons	0.22481	Pounds (force)
Pounds of water	0.1198	Gallons
Pounds per square inch	0.068046	Standard atmospheres
Pounds per square inch	51.715	Millimeters of mercury at 0°C
Standard atmospheres	14.696	Pounds per square inch
Tons (short)	2,000	Pounds
Tons (short)	0.907185	Metric tons
Tons (metric)	2,204.62	Pounds
Tons of refrigeration	12,000	Btu per hour
Therms	1×10^5	Btu
Watts	3.413	Btu per hour
Watts	0.00134	Horsepower

Glossary

absorbent—the less volatile of the two working fluids used in an absorption cooling device.

absorber—the blackened surface in a collector that absorbs the solar radiation and converts it to heat energy.

absorptance—the ratio of solar energy absorbed by a surface to the solar energy striking it.

active system—a solar heating or cooling system that requires external mechanical power to move the collected heat.

air-type collector—a collector with air as the heat transfer fluid.

altitude—the angular distance from the horizon to the sun.

ambient temperature—the temperature of surrounding outside air.

aperture—solar collection area.

ASHRAE—abbreviation for the American Society of Heating, Air-Conditioning and Refrigerating Engineers.

auxiliary heat—the extra heat provided by a conventional heating system for periods of cloudiness or intense cold, when a solar heating system cannot provide enough.

azimuth—the angular distance between true south and the point on the horizon directly below the sun.

balance of system—the components other than the photovoltaic cells that make up a photovoltaic system.

British thermal unit, or Btu—the quantity of heat needed to raise the temperature of 1 pound of water 1°F.

calorie—the quantity of heat needed to raise the temperature of 1 gram of water 1°C.

closed-loop—any loop in the system separated from other loops by a heat exchanger; often closed to the atmosphere as well.

coefficient of heat transmission, or U-value —the rate of heat loss in Btu per hour through a square foot of a wall or to the building surface when the difference between indoor and outdoor air temperatures is 1°F, measured in Btu/(hr ft^2°F).

collector—any of a wide variety of devices used to collect solar energy and covert it to heat.

collector efficiency—the ratio of heat energy extracted from a collector to the solar energy striking the cover, expressed in percent.

collector tilt—the angle between the horizontal plane and the collector plane.

concentrating collector—a device which uses reflective surfaces to concentrate the sun's rays onto a smaller area, where they are absorbed and coverted to heat energy.

conductance—a property of a slab of material equal to the quantity of heat in Btu per hour that flows through one square foot of the slab when a 1 °F temperature difference is maintained between the two sides.

conduction—the transfer of heat energy through a material by the motion of adjacent atoms and molecules.

conductivity—a measure of the ability of a material to permit conduction heat flow through it.

convection—the transfer of heat energy from one location to another by the motion of a fluid (air of liquid) which carries the heat.

cover plate—a sheet of glass or transparent plastic that sits above the absorber in a flat-plate collector.

declination—angle between the sun and the earth's tilt.

degree-day—a unit that represents a 1°F deviation from some fixed reference point (usually 65°F) in the mean daily outdoor temperature.

design heat load—the total heat loss from a house under the most severe winter conditions likely to occur.

design temperature—a temperature close to the lowest expected for a location, used to determine the design heat load.

differential controller or differential thermostat—a device that receives signals from temperature sensors, measures differences, and sends commands to pumps, fans, valves, or dampers.

diffuse radiation—sunlight that is scattered from air molecules, dust, and water vapor and comes from the entire sky vault.

direct gain—techniques of solar heating in which sunlight enters a house through the windows and is absorbed inside.

direct radiation—solar radiation that comes straight from the sun, casting clear shadows on a clear day.

domestic hot water—potable water used for washing, cooking, and cleaning.

doped—treating a substance with phosphorous (to create a negative charge) or boron (to create a positive charge).

double-glazed—covered by two panes of glass or other transparent material.

emittance—a measure of the propensity of a material to emit thermal radiation.

eutectic salts—a group of phase-change materials that melt at low temperatures, absorbing large quantities of heat.

evacuated-tube collector—a collector made of small absorbers in evacuated glass cylinders.

f-chart—computer simulation that estimates the performance of active and passive solar energy systems.

flat-plate collector—a solar collection device in which sunlight is converted to heat on a plane surface, without the aid of reflecting surfaces to concentrate the rays.

focusing collector—a collector that uses mirrors to reflect sunlight onto a small receiver.

forced convection—the transfer of heat by the flow of warm fluids, driven by fans, blowers, or pumps.

Glaubers salt—sodium sulfate (Na_2SO_4 $10H_2O$), a eutectic salt that melts at 90°F and absorbs about 104 Btu per pound as it does so.

gravity convection—the natural movement of heat through a body of fluid that occurs when a warm fluid rises and cool fluid sinks under the influence of gravity.

header—the pipe that runs across the top (or bottom) of an absorber plate, gathering (or distributing) the heat transfer fluid from (or to) the grid of pipes that run across the absorber surface.

heat capacity—a property of a material, defined as the quantity of heat needed to raise one cubic foot of the material 1°F.

heat exchanger—a device, such as a coiled copper tube immersed in a tank of water, that is used to transfer heat from one fluid to another through an intervening metal surface.

heating season—the period from about October 1 to about May 1, during which additional heat is needed to keep a house warm.

heat pump—a mechanical device that transfers heat from one medium (called the heat source) to another (the heat sink), thereby cooling the first and warming the second.

heat sink—a medium or container to which heat flows (see heat pump).

heat source—a medium or container from which heat flows (see heat pump).

heat storage—a device or medium that absorbs collected solar heat and stores it for periods of inclement or cold weather.

heat storage capacity—the ability of a material to store heat as its temperature increases.

hybrid system—a system that uses both active and passive methods to collect, distribute, and store heat.

indirect system—an active solar heating or cooling system in which the solar heat is collected exterior to the building and transferred inside using ducts or piping and, usually, fans or pumps.

infiltration—the movement of outdoor air into the interior of a building through cracks around windows and doors or in walls, roofs, and floors.

infrared radiation—electromagnetic radiation, whether from the sun or a warm body, that has wavelengths longer than visible light.

insolation—the total amount of solar radiation, direct, diffuse, and reflected, striking a surface exposed to the sky.

insulation—a material with high resistance or R-value that is used to retard heat flow.

integrated system—a passive solar heating or cooling system in which the solar heat is absorbed in the walls or roof of a dwelling and flows to the rooms without the aid of complex piping, ducts, fans, or pumps.

kilowatt—a measure of power equal to one thousand watts, approximately 1 1/3 horsepower, usually applied to electricity.

kilowatt-hour—the amount of energy equivalent to one kilowatt of power used for one hour—3,413 Btu.

langley—a measure of solar radiation, equal to one calorie per square centimeter.

latent heat—the amount of heat, in Btu, needed for a material to change phase from a liquid to a gas, or liquid to a solid, and back again.

life-cycle costing—an estimating method in which the long-term costs such as energy consumption, maintenance, and repair can be included in the comparison of several system alternatives.

liquid collector—a collector with a liquid as the heat transfer fluid.

load collector ratio (LCR)—a building's heat loss per degree day divided by the area of south glazing.

natural convection—see gravity convection.

nocturnal cooling—the cooling of a building or heat storage device by the radiation of excess heat into the night sky.

open loop—any loop in the system where potable water is used for collection and storage; open to the atmosphere as well.

parabolic collector—a concentratory collector with a parabolic-shaped reflector.

passive system—a solar heating or cooling system that uses no external mechanical power to move the collected solar heat.

percentage of possible sunshine—the percentage of daytime hours during which there is enough direct solar radiation to cast a shadow.

phase-change material—a substance that stores and releases latent heat when it changes from a liquid to a solid or gas.

photons—particles of light energy that transfer energy to electrons in photovoltaic cells.

photosynthesis—the conversion of solar energy to chemical energy by the action of chlorophyll in plants and algae.

photovoltaic cells—semi-conductor devices that convert solar energy into electricity.

present value—the value in today's dollars of something to be received at a later date.

profile angle—the angle between the horizon and the sun's rays, in a vertical plane perpendicular to a window; used to size fixed overhangs for shading.

radiant panels—panels with integral passages for the flow of warm fluids, either air or liquids. Heat from the fluid is conducted through the metal and transferred to the rooms by thermal radiation.

radiation—the flow of energy across open space via electromagnetic waves, such as visible light.

reflected radiation—sunlight that is reflected from surrounding trees, terrain, or buildings onto another surface.

refrigerant—a liquid such as Freon that is used in phase-change collectors or cooling devices to absorb heat from surrounding air or liquids as it evaporates.

resistance or R-value—the tendency of a material to retard the flow of heat, measured in (hr ft^2°F)/Btu.

retrofitting—the application of a solar heating or cooling system to an existing building.

risers—the flow channels or pipes that distribute the heat transfer liquid across the face of an absorber.

R-value—see resistance.

seasonal efficiency—the ratio of solar energy collected and used to that striking the collector, over an entire heating season.

selective surface—an absorber coating that absorbs most of the sunlight hitting it but emits very little thermal radiation.

shading coefficient—the ratio of the solar heat gain through a specific glazing system to the total solar heat gain through a single layer of clear, double-strength glass.

shading mask—a section of a circle that is characteristic of a particular shading device; this mask is superimposed on a circular sun path diagram to determine the time of day and the months of the year when a window will be shaded by the device.

solar declination—see declination.

solar house—a dwelling that obtains a large part, though not neccessarily all, of its heat from the sun.

solar load ratio (SLR)—ratio of solar gain to heat load.

solar radiation—electromagnetic radiation emitted by the sun.

solar savings fraction (SSF)—ratio of solar savings to net reference load.

specific heat—the amount of heat, in Btu, needed to raise the temperature of 1 pound of a material 1°F.

sun path diagram—a circular projection of the sky vault, similar to a map, that can be used to determine solar positions and to calculate shading.

sunspace—a glazed room on the south side of a building that collects solar energy to heat that building.

thermal capacity—the quantity of heat needed to warm a collector up to its operating temperature.

thermal mass, or thermal inertia—the tendency of a building with large quantities of heavy materials to remain at the same temperature or to fluctuate only very slowly; also, the overall heat storage capacity of a building.

thermal radiation—electromagnetic radiation emitted by a warm body.

thermosiphoning—see gravity convection.

tilt angle—the angle that a flat collector surface forms with the horizontal.

trickle-type collector—a collector in which the heat transfer liquid flows down channels in the front face of the absorber.

tube-in-plate absorber—copper sheet metal absorber plate in which the heat transfer fluid flows through tubes in the plate.

ultraviolet radiation—electromagnetic radiation, usually from the sun, with wavelengths shorter than visible light.

unglazed collector—a collector with no transparent cover plate.

U-value—see coefficient of heat transmission.

Index

Index

Index